Two-Way Radios & Scanners
FOR DUMMIES®

by H. Ward Silver
Author of *Ham Radios For Dummies*

WILEY

Wiley Publishing, Inc.

Two-Way Radios & Scanners For Dummies®

Published by
Wiley Publishing, Inc.
111 River Street
Hoboken, NJ 07030-5774

www.wiley.com

Copyright © 2005 by Wiley Publishing, Inc., Indianapolis, Indiana

Published by Wiley Publishing, Inc., Indianapolis, Indiana

Published simultaneously in Canada

For general information on our other products and services, please contact our Customer Care Department within the U.S. at 800-762-2974, outside the U.S. at 317-572-3993, or fax 317-572-4002.

For technical support, please visit www.wiley.com/techsupport.

Wiley also publishes its books in a variety of electronic formats. Some content that appears in print may not be available in electronic books.

Library of Congress Control Number: 2005924591

ISBN-13: 978-0-7645-9582-0

ISBN-10: 0-7645-9582-2

10 9 8 7 6 5 4 3 2 1

1O/RS/QX/QV/IN

WILEY

About the Author

H. Ward Silver has experienced a 20-year career as an electrical engineer developing instrumentation and medical electronics. He also spent 8 years in broadcasting, both programming and engineering. In 2000, he turned to teaching and writing as a second career, producing *Ham Radios For Dummies* in 2004. He supports Seattle University's Electrical and Computer Engineering Department in laboratory instruction. He is an avid Amateur Radio operator, Extra Class, first licensed in 1972. Each month, his columns and articles can be found in the national ham radio magazine, *QST,* published by the American Radio Relay League (ARRL). He is the author of the ARRL's online courses in Antenna Design and Construction, Analog Electronics, and Digital Electronics. When not in front of a computer screen, you will find him working on his mandolin technique and compositions.

Dedication

This book is dedicated to my mom, who enabled her mad scientist son to conduct his basement activities in radio and science. Also to our family cat, Mirage, who carefully supervised the author throughout the creation of this book.

Author's Acknowledgments

I would like to gratefully acknowledge the Uniden America Corporation for providing equipment, a number of excellent graphics, and for their technical reviews of several chapters. Paul Opitz, Uniden Marketing Manager, was particularly generous with his time and assistance. I must also recognize the professional contributions of my editor, Nicole Haims, who patiently trimmed and guided each chapter to the polished form you enjoy. My technical editor, Kirk Kleinschmidt, also provided invaluable review and commentary. Thanks also to Melody Layne at Wiley for enabling me to write this second title.

Much gratitude also goes to the members of the Vashon-Maury Island Radio Club and of the Vashon Disaster Preparedness Coalition, who were the guinea pigs for many of my planning and training experiments.

Thank you!

Publisher's Acknowledgments

We're proud of this book; please send us your comments through our online registration form located at www.dummies.com/register/.

Some of the people who helped bring this book to market include the following:

Acquisitions, Editorial,
and Media Development

Project Editor: Nicole Haims

Acquisitions Editor: Melody Layne

Technical Editor: Kirk Kleinschmidt

Editorial Manager: Carol Sheehan

Permissions Editor: Laura Moss

Media Development Manager:
Laura VanWinkle

Media Development Supervisor:
Richard Graves

Editorial Assistant: Amanda Foxworth

Cartoons: Rich Tennant (www.the5thwave.com)

Composition Services

Project Coordinator: Maridee Ennis

Layout and Graphics: Carl Byers, Andrea Dahl, Kelly Emkow, Joyce Haughey, Mary Gillot Virgin, Lynsey Osborn

Proofreaders: Leeann Harney, Jessica Kramer, Joe Niesen, TECHBOOKS Production Services

Indexer: TECHBOOKS Production Services

Publishing and Editorial for Technology Dummies

    Richard Swadley, Vice President and Executive Group Publisher

    Andy Cummings, Vice President and Publisher

    Mary Bednarek, Executive Acquisitions Director

    Mary C. Corder, Editorial Director

Publishing for Consumer Dummies

    Diane Graves Steele, Vice President and Publisher

    Joyce Pepple, Acquisitions Director

Composition Services

    Gerry Fahey, Vice President of Production Services

    Debbie Stailey, Director of Composition Services

Contents at a Glance

Table of Contents

Introduction

*T*o all my readers, welcome to the amazing world of radio that awaits you just behind the front panel power switch. There's a lot more radio out there than you might think! This book helps introduce you to it.

If you've just become interested in radio, *Two-Way Radios & Scanners For Dummies* contains plenty of information to explain what radio is all about. You'll have more fun and success by mastering the basics, and so will the friends, family, and associates with whom you're trying to communicate. This book answers a lot of questions about the jargon being thrown around by manuals and by other radio aficionados. It's the perfect companion for people who want to maximize their radio experience and utility. Whether you're using your radio for work or for recreation, whether you already own a radio or are still browsing for the right system, this book gives you all you need to know in one convenient package.

About This Book

This book is designed with two reading audiences in mind:

- **New users:** New to radio? Welcome! This book helps you become a confident, knowledgeable radio user. All the new activities, particularly the technical ones, can be intimidating to newcomers. If you're still shopping for a radio, I give you some tips on what to look for in a good radio. If you just bought a radio, this book gives you a tour of the main features. By keeping *Two-Way Radios & Scanners For Dummies* handy, you will have an easier time getting your radio or scanner connected and working right. I cover the basics of properly putting together a station and give you the fundamentals of on-the-air behavior. Use this book as your personal radio buddy and soon you'll be sounding (and listening) like a pro!

- **Radio users who want more information:** Those of you already using a radio can find new ways of putting that radio to work and get a better understanding of how the radio does what it does. You can find out about other types of radios and services, as well. I give you lots of resources to help you get more out of the radios you have and make good choices about antennas, accessories, and new radios. There is a little something for every level of operator!

What You're Not to Read

If you just got a radio, I'll bet you're wondering, "What do I do now?" New radio users need a helping hand to find out how to use it properly. *Two-Way Radios & Scanners For Dummies* will be there to look over your shoulder until you are confidently making good use of your equipment. I make the basics of radio available in the first parts of the book so you won't have to be a radio whiz to assemble and use your equipment. If you already know these basics, you can skip Part I and move on to the chapters that cover specific radio services and functions.

As you make your way through *Two-Way Radios & Scanners For Dummies,* feel free to skip around to whatever areas your interests take you. Within a chapter, you can ignore the sidebars, if you like, which contain extra-technical details or slightly off-the-beaten-path information. The book is written so that you can open it anywhere at any time.

Foolish Assumptions

The first thing I assume is that you either have a radio or will soon be purchasing one. Maybe you have fiddled around with a radio that belongs to a friend and want to know a bit more about what types of radios and scanners are available before you take the plunge. This book helps you select a radio based on your needs; to get the most out of the book, you'll need to borrow or buy your own radio so that you can tinker around.

If you're just getting started, the amount of information available about radios, radio technology, electronics, and more can be overwhelming! You can start relaxing right now. I assume that you don't know a single thing about radio technology, so you can enjoy *Two-Way Radios & Scanners For Dummies.* Every chapter offers supporting explanations of radio jargon and conventions in plain English.

I assume that you know how to use a computer and that you have Internet access to use the online resources I provide. Internet access is also useful for finding expanded explanations of more complex topics related to radio technology.

If you need to obtain a license to use your type of radio, *Two-Way Radios & Scanners For Dummies* assumes that you need information about and guidance through the licensing process.

How This Book Is Organized

Two-Way Radios & Scanners For Dummies is divided into parts to meet the needs of beginners, dabblers, and radio enthusiasts.

Part I: Making Radio a Hobby, a Habit, or a Helper

If you don't have much information about radios but you know enough that you're interested in using them, or if you've just obtained a radio, this is where you start reading. Part I gives you the big picture of how radio technology works. Find out the differences between the various radios and radio communications. Because some of the radios require a license to use, this part also covers what you need to do to get the required paperwork in place.

Part II: Two-Way Radios at Home, Work, and Play

Radios that connect people are more popular than ever and there are so many types! Just about every activity you can think of has its own specialized radio service and Part II gives you information about the most popular services around. It starts with a discussion about the handy palm-sized units for short-range person-to-person chatting and moves on to investigate Citizens Band radio, emergency communications, and using radios for your business. You can also find out about radios used for boating. The part concludes with a chapter on the most powerful personal radio service of all, amateur radio. If you already have a radio, this part shows you how to use it. If you're trying to choose a radio to fill a specific need, this is where you get the information needed to decide.

Part III: Listening In: Scanning and Shortwave Listening

If you're into radios (which you are); then you'll enjoy finding out about high-performance receivers that pull in signals from across town *(scanners)* and around the world *(shortwave receivers)*. Part III discusses what distinguishes the various types of listen-only radios from each other and helps you choose the right technology. It offers several popular ways of using these special radios.

Part IV: Getting Technical with Your Radio

After Parts II and III, your appetite will be whetted to get busy putting together a complete radio station. While you could just rush out and lash it all together, you'll have *much* better results if you set everything up the right way first! Part IV shows you how to do a proper installation, whether at home or on the road. It offers helpful information to guide you in choosing the right accessories and batteries. Many radios can interact with computers, so I cover setting up an interface between your PC and your radio, too. Finally, the reality is that occasionally you'll have to do some troubleshooting. This bothersome, but inevitable task is much easier if you master the fundamentals.

Part V: The Part of Tens

A familiar facet to all the books in the *For Dummies* books is the Part of Tens, which consists of several condensed lists that provide some really helpful and hopefully memorable ideas. The chapters in Part V reinforce and support the extended discussions of the earlier chapters to guide you in your use of a radio. This part offers tips, secrets, radio first-aid goodies, and ideas for enjoyable radio operation.

Appendix

The appendix contains a glossary that is chock-full of terms and definitions for easy accessibility and use. Look here to find an abbreviation or check out a technical fine point.

Conventions Used in This Book

To make the reading experience as clear and uncluttered as possible, a consistent presentation style is used.

- *Italics* are used to note a new or important term, such as *ham radio*.
- I use italics and lowercase when I recommend search terms for further investigation online, like this: "Try entering *citizens band club [your hometown]* in an Internet search engine to find like-minded individuals near you."
- Web site URLs (addresses) are designed by using a `monospace` font. Hyperlinks are <u>underlined</u>.

You see the following icons used as markers for special types of information:

The Tip icon alerts you to a hint that will help you understand a technical or operating topic. Sometimes radio mavens refer to tips as _hints and kinks._

This icon signals a useful, but geekily technical radio fact. You might not need to know the information contained here, but if you're a closet (or not so closet) radio dork, you will find the information both interesting and handy.

Whenever a common problem or potential "oops" arises, this icon is nearby. Before you become experienced, it's easy to get hung up on some of the little things, so this icon helps you out if you get stuck.

This icon lets you know that there are serious legal, safety, or performance issues associated with the topic of discussion. Watch for this icon to avoid damaging your radio, getting harmed by electricity, or getting yourself into some other serious hot water along the way.

This icon indicates information that you need to know — if not now, soon. Store this information away for that moment (it will come) when you need to know it and you'll have more fun on the air.

Where to Go From Here

If you are just beginning to find out about using two-way radios or scanners, I recommend that you read through Part I before proceeding. For those of you who are more experienced with radio and are looking for in-depth information, you'll probably make the best use of this book by jumping around from chapter to chapter as your interests carry you. You can find the right chapter by browsing through the table of contents, index, or just skimming the headings along the way.

All readers will benefit from cruising through the Appendix for helpful definitions of common radio terms.

Will radio go the way of the rotary phone, bypassed by high-speed digital technologies? I don't believe so. Its personal utility and innate ability to be used directly without requiring the support of communications networks virtually guarantee it a place in the communications pantheon. Invented more than 100 years ago, radio continues to serve us faithfully and will for the foreseeable future.

Part I
Making Radio a Habit, a Hobby, or a Helper

The 5th Wave   By Rich Tennant

"Yes, it's wireless, and yes, it weighs less than a pound, and yes, it has multiuser functionality... but it's a stapler."

In This Part . . .

Welcome to the world of two-way radios! Whether you plan on talking on a radio or just listening, radio technology is a fascinating medium that has long captured the imagination of folks just like you. The more you know about the technology of radios, the more you'll understand what's happening on the airwaves.

This part contains chapters that introduce you to two-way radios and explain some of the technology that makes them work. I begin with a general overview of the modern radio. I also cover some technological basics. Armed with that knowledge, you can discover in Part I the different ways you can put radio to work in your daily life. Because some radios require a license, I introduce you to the Federal Communications Commission. You must follow the FCC's guidelines, and you may need to use its licensing system, depending on what kind of radio service you use. You'll be a budding radio maven by the time you finish reading these chapters.

Chapter 1

Introducing Radios and the Wireless World

*1*f you go to the local electronics emporium, I bet you will see a display labeled Two-Way Radios. There you'll find the radios about which this book is written — *handheld radios* that use the unlicensed Family Radio Service (FRS) and their big brothers in the General Mobile Radio Service (GMRS). You may see some ham radios or maybe radios that use the *dot* frequencies. By talking as well as listening, these *two-way* radios connect people together when they talk business, participate in recreational activities, or need to communicate in emergencies.

A little farther down the aisle you'll meet the *one-way radios*, scanners and communications receivers — devices that let you listen in without talking back. *Scanning* the airwaves has become quite popular. You can follow the activities of your local public safety agencies, monitor business users, and listen in at sporting events just by punching a few buttons. *Shortwave radio* is going strong, as well, with broadcasts from around the globe arriving at every hour of the day and night. With a simple radio and minimal antenna, you can get connected in a way that the Internet just can't match.

Read on as I fill in the blanks for common terms and ideas that make radio the valuable resource it is. Consider this a book about grown-up radio!

Understanding How Radios Fit into a Wireless World

You hear people bandy about the term *wireless* a lot these days. Everything seems to have a wireless option — wireless toasters, wireless toys, wireless tote bags! All of them make use of radio technology to communicate. The days may be gone when radios were the only wireless game in town, but nowadays radios are everywhere!

Radios, PCs, and phones — Oh, my!

What is the most common two-way radio on the planet? Hint: You've probably used one without even thinking of it as a radio. It's the mobile phone! Those ubiquitous little items glued to everyone's ear contain a sophisticated two-way radio that sends and receives phone calls in a never-ending stream of signals. In fact, you have probably seen people using two radios at once, typing with one hand on a laptop connected to a wireless network at the coffee shop or mall while holding a mobile phone with the other. Both devices are examples of specialized radios. Figure 1-1 shows some more.

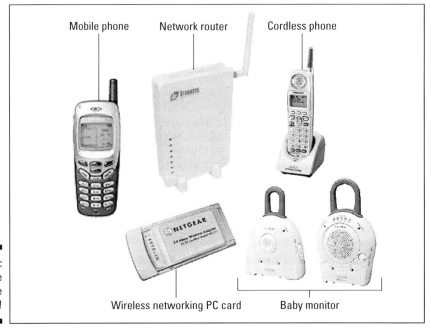

Mobile phone · Network router · Cordless phone · Wireless networking PC card · Baby monitor

Figure 1-1:
All these devices are radios!

Even many of the ID tags attached to goods you purchase at the store are very simple radios. So where does the regular, old-fashioned radio stop and the new, fancy radio begin? In this book, when I talk about a *radio*, I mean the traditional form that allows you to listen to signals from broadcasters or other radio users, as well as those that can transmit to communicate directly with someone else who also has a radio.

Why get enthusiastic about radio?

With all this communications gadgetry, why would you need a radio, anyway? Isn't radio obsolete? After all, you're just a few keystrokes away from the Internet via wireless networking. Flip open your mobile phone and reach just about anyone, just about anywhere on the earth's surface in seconds. Even considering the latest technology, two-way radio can do things that other fancy techno-gadgets can't.

What happens when you try to make a call and you see a message on your phone that says No Service? What happens when you try to access the Internet with a wireless connection and you see a message on-screen that says No Connection? People on the go, whether they're just roaming around a mall or driving through the Rocky Mountains, may find that a two-way radio does a better job at facilitating communication (at a lower cost, mind you) than these commercial service options. Why use up those calling plan minutes when a pocket-sized radio can keep you in touch with your family elsewhere at the ball game? If you're stuck on the road between cell towers, a *Citizens Band* (CB) radio is nice to have if you need to call for help. Radio gives you lots of alternative ways to communicate.

Plus, there's the fun part. You're doing it yourself and not relying on some big company (or paying their monthly bills) to communicate. After you become familiar with your radio, you might find yourself taking pride in seeing how far you can get. Or maybe you can take pride in being the first to respond to a request for help. Even when you're not transmitting yourself, you can have hours of enjoyment just listening to the public safety services around town with a scanner or pulling in the morning news from halfway around the globe on a shortwave radio.

Communicating person to person

Two-way radios are used the world over to put people in touch with people. Radio does this with a bare minimum of fancy supporting technology and so provides a very inexpensive way of communicating. Examples of person-to-person radios shown in Figure 1-2 that you encounter in this book include

✔ Handheld *FRS/GMRS* radios (Chapter 4)

✔ *Citizens Band* (or CB) radios (Chapter 5)

✔ Business and public safety radio systems (Chapter 7)

✔ Marine radios for sailors (Chapter 8)

✔ *Ham radio,* the most powerful personal radio of all (Chapter 9)

You also find how to use a scanner to listen in as others use their own radios.

CB radio

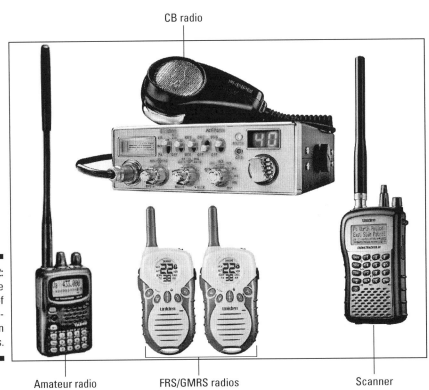

Figure 1-2: Some examples of person-to-person radios.

Amateur radio   FRS/GMRS radios   Scanner

Images courtesy of Uniden Corporation

Saying, "Mayday!"

Marine radio, Citizens Band, amateur (or ham) radio — these services all have designated channels or common frequencies where help can be a transmission away. And yes, you say, "Mayday!" just like they do in the movies. Personally, I've reported a number of automobile accidents and stranded motorists to the 911 dispatchers using my ham radio. I've also monitored Marine Channel 16 as a grounded boater called looking for help from the Coast Guard. The most important emergency communications in which I've ever participated occurred during 1991, when troops from the Soviet Union surrounded the Lithuanian parliament building. Alerted by a fellow ham, I listened as the ham radio club station inside the parliament building relayed reports to a local Vilnius station, which then relayed them to a California station and on to the U.S. State Department. As the hours passed, we helped keep the frequency clear and supported a new American station in New Hampshire to take over when conditions to the West Coast began to fade. Later on, when we could only hear the new European relay station in Belgium, we were about to sign off when a formal message came through transferring the Lithuanian government's power of authority to a cabinet minister, who was kept outside the country for that very purpose! That was possibly one of the most important emergency messages ever sent by amateur radio. Since then, I've learned to keep my ears open; you never know what the airwaves will bring.

Communicating in an emergency

Radio really shows its worth when the chips are down (and so are the mobile phones and Internet systems). Not many people realize that commercial communications systems (such as *POTS* — or plain old telephone service) are designed to handle only a few percent of their subscribers at any given time. In a smaller-scale emergency (such as a blizzard), you are likely to find the systems unavailable due to overload. This lack of service can last for many hours and in a true disaster (such as an earthquake, tornado, or hurricane), hours, or even days, may pass before service begins to be restored. In times like these, having a radio that doesn't rely on a separate company or special systems can literally be a lifesaver.

In an emergency, not only do you need to know how to make the best use your radio, and where to find help, but you also need to know how to interact with others in a similar situation, and with the people providing assistance. Chapter 6 leads you through the basics of preparing yourself and your radio for an emergency. I also discuss techniques for communicating efficiently under difficult circumstances, and provide you with resources for training and learning about emergency communication, or *emcomm*.

Using your radio for fun

You may have purchased a radio for pure utility, but (along with millions of other enthusiasts) you may find yourself enjoying making the radio work better and improving your own skills. For example, mobile CB and marine radio enthusiasts have many opportunities to experiment with antenna style and placement to get the most range on the road and on the water.

Using a radio can enhance the experiences of other activities. For example, auto racing fans use scanners to listen in to their favorite drivers talk with their pit crews. Hikers, campers, and recreational vehicle (RV) users often use radios to keep in touch with others and as safety aids. The worldwide community of folks that combine outdoors skills and radios in orienteering and direction-finding competitions is sizeable. Hang gliders and hot air balloonists can often be heard coordinating their flights with handheld radios.

Ham radio licensees have the most latitude to experiment and practice than users of any other type of radio. After passing the required test and receiving his or her Federal Communications Commission (FCC) license, *hams* (the term affectionately used to describe ham radio users) have access to a huge range of frequencies, can build and modify their own radios, use them for all sorts of activities, including public service, and even operate their radios in foreign countries. Think of hams as having graduated from Radio University and you've got the picture.

Putting radios to work

Public safety and service workers, such as fire and police officers, paramedics, and transportation and wildlife officers, all make use of radio on a daily basis. Radios like those in Figure 1-3 are commonly used by security staff at public events and concerts. To pilots, radio is a lifeline in the sky, guiding them safely from point to point and keeping them informed about conditions ahead and on the ground.

If you use a radio as part of your job, you can get the most performance out of your radio (and, as a result, do your job more effectively) when you understand how it works. You also raise your understanding of the radio system and interact at a much higher level with radio system planners and technicians.

Radio technology is a key element to making the world's extensive aviation industry fly, so to speak. Pilots, air-traffic controllers, and airport managers all engage in an intricate choreography that ensures our safety in the air and on the ground. The dance is conducted over the radio waves, and you can listen in any time. If you're an aspiring pilot, keeping an ear on the aviation channels is a great way to improve your own skills so that you feel more comfortable in the pilot's seat.

Aviation cockpit radio | Handheld public safety radio

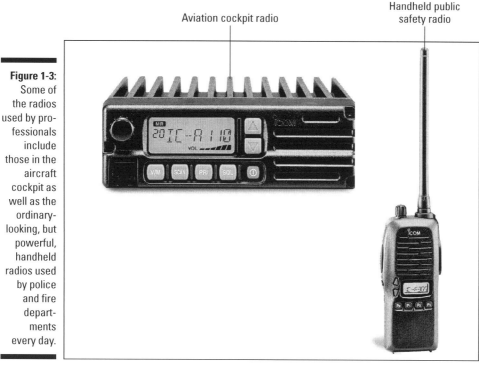

Figure 1-3:
Some of the radios used by professionals include those in the aircraft cockpit as well as the ordinary-looking, but powerful, handheld radios used by police and fire departments every day.

Images courtesy of ICOM America Corporation

Listening to the radio transmissions made by professionals in the course of their work is fascinating — just ask any scanner enthusiast! News organizations make no secret of the fact that they have several scanners hot on the trail of developing stories, 24 hours a day. By tuning in yourself, you have the same access to breaking news as the local broadcast pros.

Even if you only want to use a radio to get the job done, finding out how the radio does its job helps you pick the right radio with the right signal. You can plan more accurately, train more quickly, and make use of your investment more efficiently.

Introducing Radio's Unique (And Magical) Forms

Radio signals are whizzing around the planet, between airplanes, up and down to satellites, and off into space. This is happening all the time and the only thing you need to get in on the action is (surprise!) the right radio. That and a little bit of helpful information and instruction (surprise!) provided in this book to launch you beyond the basic functions. You'll be amazed at how much even a simple radio can do if used by a knowledgeable owner — that's you.

Most of us use a radio as just another appliance — you just turn it on and get a specific radio station or program. Your car radio has a handful of buttons that can be set to individual stations, but what's *between* those stations? That little handheld radio you bought may have more than a dozen channels. What goes on there? The keys to finding out lay buried in your owner's manual. I can help you unravel the jargon and explain the purpose of common controls and settings so that you can play your radio like a musical instrument.

In Chapter 2, I discuss some basics on what makes radio work, but for now you should know that the signals have a wide range of characteristics. Some radio signals can travel, or *propagate,* for extremely long distances with a proper antenna. Often, signals are reflected off the upper layers of the atmosphere and return to the ground hundreds or thousands of miles away. Other signals ignore these layers, blasting off directly to or from a satellite. Sometimes, the weather gets into the act and allows signals that normally wouldn't travel beyond the horizon to be received in the next state. This variability leads to exciting adventures, tracking the wily signals from points known and unknown.

Not only do the signals travel surprising distances, but also there are many different types of radio signals. Some transmissions are of voices in broadcasts or conversations. Others are designed for carrying data or pictures that you can display or record. When you know where to listen, you can also hear the venerable Morse code and the almost-as-old *radioteletype,* or RTTY (pronounced "ritty" by those in the know).

You can use your computer to decode and display most of these signals with free software downloadable from the Internet. Get your weather map directly from a satellite without waiting for the TV news or logging on to the Web!

If you start viewing your radio as another set of eyes and ears, you'll discover that the number of signals at your fingertips, weak and strong, is enormous. It's the same as finding how your car works. Driving takes on a whole new dimension when you know what the car is doing and you become a much better driver and owner, as well. There's a whole 'nother world waiting for you behind a radio's power switch!

What You Can Do with a Radio

You probably had a very specific need in mind when you first starting looking for a radio. Maybe you've been able to satisfy that initial need, but if you're reading this book, you are likely the type of person who is interested in getting the most out of the things that you use. You've probably seen radios used in many common ways or *applications.* The following sections suggest a few applications for your radio that you may not have considered. Maybe the information here will spark a few new ideas.

Roger: Sharing information

A two-way radio's primary purpose is to connect people so that they can talk to each other. Usually, those people want to be connected because they are sharing information. Here are some good examples of how radio can enhance everyday info-sharing activities:

- ✔ **Becoming a severe weather watcher:** The SKYWARN program run by the National Oceanic and Atmospheric Administration (NOAA) relies on volunteers just like you to spot developing weather systems and relay your observations via radio to the National Weather Service. Similar groups, such as the British Storm Watcher's Community (`groups.msn.com/BritishStormWatchersCommunity/welcome.msnw`) and the German Severe Weather group (`www.germansevereweather.de/eindex.html`) exist in other countries.

- ✔ **Coordinating a school, church, or community event:** Keeping volunteers organized is so much easier with radio. After you run a parade, carnival, or outing with everyone on the radio, you'll wonder how you ever managed without radios.

- ✔ **Organizing your neighborhood for block monitoring or disaster relief:** In an emergency that eliminates phone services, being able to maintain contact with your neighbors without having to physically travel to each house is tremendously reassuring. Radio service is an excellent way to help each other.

- ✔ **Teaching your kids:** Giving kids a secret radio link with a friend is wonderful. There's nothing better than having your very own private channel. A few hours doing chores can provide an inexpensive pair of radios just right for best friends. What starts as just a fun gadget can lead to a lifelong interest in the real technology behind the buttons. A kid that can use a radio can make a real contribution at school, church, or community functions, as well!

Using your radio at work

When you begin using radios for business, you must abide by certain rules regarding which radios and channels are available. Nevertheless, you can be quite creative with radio at very low cost. Here are some options for you:

✔ **Keeping tabs on delivery vehicles or drivers:** The Citizens Band services are designed just for this type of use. An inexpensive radio with a cigarette lighter plug and a clip-on antenna turns a part-time delivery person's car into a radio-linked business office without the expense of mobile phones.

✔ **Coordinating marina operations:** By using a combination of handheld short-range and regular marine channel radios, a *harbor master* can coordinate incoming boaters, keep tabs on those out on the lake or river, communicate with marina staff, and talk to the marina store or fueling station.

✔ **Setting up a campground intercom:** With campers spread out over many acres, keeping order can be quite a chore for a campground manager or supervisor. If you hang out a sign directing campers to use a specific radio channel, they instantly have a way of communicating with you and each other.

✔ **Coordinating the emergency communications plan for your business:** Your employees probably use radios for day-to-day operations, but do they know how to use them in an emergency, such as a fire? Knowing where everyone is (and isn't) is a huge relief and helps avoid panic.

Listening in with a scanner

Keeping an ear on the local public safety action is the most popular use of a special radio receiver, commonly known as a *scanner*. There are quite a number of other ways to make good use of these receivers:

✔ **Monitoring the action at auto races or air shows:** Chapter 14 shows you how to become part of the team, allowing you to listen in on the pit chatter, adding a lot to your enjoyment of the event. This is a great way to discover how the drivers, pilots, and players do their thing and how the events are organized.

✔ **Picking the right line:** If you ride ferries or spend time in traffic lanes, having a portable or mobile scanner can let you listen directly to the transportation workers. You can find out what lane is held up and why, when the next boat will arrive, if a new lane is about to open up — wouldn't you just love to know?

✔ **Getting your weather directly from the satellite:** You don't have to have a 30-foot dish to pick up signals from the low-orbit weather satellites (called *birds* by the initiated). All it takes is a small antenna, a VHF receiver (VHF stands for *very high frequency*), and a computer to automatically record all types of photos and data from the many *remote sensing* satellites.

✔ **Tuning in for military surprises:** Although the men and women in uniform don't publish their lists of frequencies, you can make a pretty good guess about where to listen in when the military is active. Depending on their location, you might hear pilots talking to each other or to their command center from thousands of miles away!

Chasing broadcasts

You may have listened to one of the major shortwave broadcast services, such as the BBC or Deutsche Welle. While on the road, you may have heard the fluttery sound of a faraway AM (amplitude modulation) radio station coming in at night. (See Chapter 2 for more information on AM and other terms.) If those wet your whistle, here are a few more ideas:

✔ **Finding news in your native language when you're traveling abroad:** If you're not fluent in the local lingo, you'll find yourself yearning for news from home. A portable receiver and a hunk of wire can open a channel that receives shortwave broadcasts from anywhere on earth.

✔ **Crossing borders:** You can practice rusty language skills or experience entirely new holidays and festivals through the magic of shortwave radio. You find an immediacy and presence on shortwaves that the Internet just can't match.

✔ **Having a distance (DX) contest:** Do you have friends with radios of their own? Challenge them to see who can receive the most distant station! Then see who will be the first to receive a confirmation or *QSL card* bearing those exotic stamps from a faraway country.

DX is an old telegraphic abbreviation that indicates *distant station*. The term is still commonly used by shortwave and amateur radio enthusiasts as shorthand for a signal received from a distant station. A *QSL card* is a hard-copy verification of reception or of the contact; it contains important details about the date and time of the contact, as well as the frequency used and signal strength. The abbreviation QSL is one of many *Q-Signals* used by radio operators (www.ac6v.com/Qsignals.htm) to abbreviate common phrases and questions.

Knowing Radio Rules and Regulations

This book is primarily written with the U.S. reader in mind, so when I talk about the musts and mustn'ts, I am thinking specifically of the Federal Communications Commission (FCC). You can find out more about the FCC and all the radio rules at the FCC's Web site, www.fcc.gov. If you're an international reader, consult the information supplied by the manufacturer of your radio and read magazines and Web sites originated in your home country for the appropriate local information.

Check the owner's manual first to find references to applicable rules in your country. The manual also provides sources so that you can find detailed regulations.

Some two-way radios, such as those that use the GMRS (or *General Mobile Radio Service*) band, require a license to operate. Whenever I discuss such radios in this book, I'll be sure to let you know right up front that a license is needed and provide you with instructions on how to file the necessary paperwork to get licensed.

Don't think that you can get away without obtaining a license. You may last for a while, but getting caught can result in the loss of your equipment, a fine, and at the least a big hassle. And for what? There's nothing inherently difficult about obtaining a license to begin with. In fact, the perks are pretty good.

Even *receive-only* radios, such as scanners and communications receivers, have a few rules to play by. For example, listening to mobile phone conversations is illegal, period, as is modifying a radio to do so. Divulging what you overhear on the air may also be illegal. Various states also have laws restricting the use of mobile scanners that can receive law enforcement frequencies. Check your local laws before putting that scanner in the car.

Getting Training (If You Need To)

After you dive into radio in a big way, you'll soon discover that you need a lot more information than any single book can provide — even this one! Luckily, there are lots of resources out there for the aspiring radio guru. The following sections give you some resources.

Books and videos

The most common resources for further radio training are print books and video programs that you can buy, rent, or borrow. Libraries often have extensive collections of training materials. Radio manufacturers want you to have the best possible experience with their products, so they often provide free operating guides, tips, and articles on their Web sites or through their dealers. The larger radio clubs and users groups often list a number of resources on their Web sites for you to try. Here are a few:

- Strong Signals (www.strongsignals.net)
- National Citizen's Band Center (www.bearcat1.com/ordercb.htm)
- American Radio Relay League (www.arrl.org)
- North American Shortwave Association (www.anarc.org/naswa)

For all the activities in this book there are numerous online communities, such as those at Yahoo! Groups (groups.yahoo.com) and MSN Groups (groups.msn.com). You can also find many e-mail lists, such as those at www.dxzone.com. Take advantage of them to get more enjoyment and utility out of your radios!

Online training

Online tutorials and training courses are getting more popular every year, and they are widely available. Quality runs from cursory overviews to in-depth professional-level certification programs. As with books and videos, check the manufacturer and organization Web sites for references or sponsorship of courses. If you're interested in *emcomm,* or emergency communication, local organizations may sponsor training courses for a low fee — or perhaps for no charge at all.

In-person training

I can't think of a better substitute for getting the assistance of a more experienced radio user. If you're fortunate enough to have such a friend or acquaintance, congratulations! Not everybody has a radio mentor (or *Elmer*) to help them over the bumpy spots. To find these folks, look for a local or regional club that covers your area of interest, such as scanning or direction finding. Regional clubs may even have a special officer whose job is to welcome newcomers and help answer questions. In general, experienced radio gurus are quite willing to help even a rank newcomer. After all, even the saltiest veterans were once in your shoes!

You, Too, Can Build and Fix Your Own Radio

You can't do much with a cellphone beyond open it, use it, and pay the bill every month. Most types of radio, though, present a golden opportunity to optimize and expand your capabilities. You can find out about electronics in a big hurry by experimenting with radio, plus you'll meet a lot of like-minded folks that enjoy getting the most out of their radios.

Limitations on opening the hood

Not quite so fast, grasshopper. Although you may have an itchy screwdriver, you need to know and heed some rules of the road. Two-way radios require

FCC *type acceptance*, which means that the manufacturer guarantees the radio will play by the rules of its service. If the radio needs to be repaired, either the manufacturer or a certified technician must repair it to restore that guarantee of proper operation. If you look at the manual or manufacturer's nameplate of your radio, you will see an *FCC ID*, which is the stamp of approval for that radio. Every radio sold in the United States has to have one.

Because most users of radios are relatively untrained, tinkering with the radios can (and does) lead to serious radio misbehavior and interference to other users. This sort of thing attracts the generally unwelcome attention of the FCC. If you don't know what you're doing, enlist the services of a technician with the appropriate certifications.

The operating manuals for most radios are available online at little or no cost, either from the manufacturer's Web site or from sites like www.w7fg. com. Try entering your radio's model number along with the word *manual* into an Internet search engine.

Fiddling about inside can also void your radio's warranty, so exercise some caution. In Part IV, I present some technical chapters on tools, installing radios, and troubleshooting. Consult these and some training and reference resources before you open up that radio.

Kits and homebrewing

If you really want to build and tinker with radio circuits, you should look into getting an amateur or ham radio license. Ham radio is a great hobby for electronic tinkerers and builders! Right behind hams come the scanner and broadcast listeners who strive to pull out weak transmissions with a collection of sensitive antennas and signal-enhancing electronics.

Both ham radio operators and the listen-only folks have lots of opportunities to build and test their own radios and accessories. A few are shown in Figure 1-4.

Accessorizing

Don't let my cautionary advice take all the fun out of playing with your radio. For most services, you can add all sorts of accessories like headsets, special microphones, signal monitors, and audio processors. Radios can often make use of dozens of different types of antennas. Instead of working on the radio itself, you can build the *station*. Installing and using accessories can help you gain a whole lot of radio savvy. Just take a look at a catalog from a marine radio dealer, for example, and you'll see just how much there is to choose from.

Ham radio operators have to pass a somewhat-technical test and so are presumed to have the technical qualifications (and the responsibility) to keep their transmitted signals up to snuff. Radio listeners don't transmit at all and there are few restrictions on what they can use to receive signals. You can build your own inexpensive shortwave broadcast receiver in a few hours!

You can find literally hundreds of different kits to create everything from power sources and test instruments to radios and antennas that push the state of the art. Along with kits, many magazines and Web sites publish circuits and instructions for builders. Building from scratch (individual components or stock materials) is called *homebrewing*, and it is a real badge of honor to use homebrew equipment and antennas.

Shortwave
receiver   Radio power meter   Homemade antenna

Figure 1-4:
Homebrew
kits include
several
gadgets that
you put
together
yourself.

Chapter 2

Discovering the Art and Science of Radio

This chapter introduces you to the fundamentals of radio. Radio, reader. Reader, radio. All set? If you're completely new to radio, this is a good chapter to have bookmarked as you read the rest of the book. You can read through it and then refer back to the relevant section as you encounter unfamiliar terms and ideas later on. If you're a more experienced radio user, use this chapter as refresher.

Doing the Wave: How Radio Waves Work

Radio signals are transmitted as a combination of electrical and magnetic waves. As I describe in the next section, you can measure radio waves by their *frequencies* or by their *wavelengths*. A radio is really just a device that measures radio signals; it's calibrated so that you can tune to a specific frequency or wavelength. The following sections explain radio signals with these two important concepts in mind.

Introducing frequency

The single most important idea to get straight is that of *frequency*, a word that occurs, um, frequently in everyday speech with many meanings. In a book about radio technology, though, frequency has a very specific meaning. It refers to the number of times every second that a radio signal changes its electrical orientation.

Radio signals are similar to sound waves in that you experience them as they move by. The signals cause *vibrations* in either your ear (as when you hear a song on the radio) or your radio's receiving circuitry (as measured by your radio). Just as sound waves cause the air to vibrate, so do radio waves cause a traveling electrical vibration.

For a more thorough description of waves and all sorts of topics associated with radio, you can use the free online encyclopedia, Wikipedia, at `en.wikipedia.org`. Just enter *radio* into the search window or browse to `en.wikipedia.org/wiki/Radio` and dive in.

You can get an idea of what frequency is all about with a piece of string or rope that's several yards long. Tie one end to a fixed object, like a tree or railing, and hold the other in your hand. Stretch the string until it's no longer loose, but do not hold it so that it's taut. Now use your free hand to pluck the string so that you generate a ripple that travels to the other end. That ripple is a wave. If you look closely, you can see that the wave consists of a big ripple in one direction, followed by a smaller ripple in the other direction. If you stand halfway along the string and let somebody else pluck it, you may see that as the wave shoots by you, the string first moves in one direction, and then moves in the opposite direction. That changing of direction is just what happens in a radio wave, but electrically. If you were a radio antenna, you would feel an electrical pull in one direction and then in the opposite direction as the radio wave passes.

Frequency describes how many times per second the radio wave reverses its electrical direction, or *polarity*, as it passes by an antenna or some other point in a radio circuit. Each complete set of reversals — one way, and then the other — is called a *cycle*.

Measuring frequency across the spectrum

Frequency is measured in cycles per second and the name for one cycle per second is a *hertz*, which is abbreviated *Hz*.

Heinrich Hertz was a 19th-century German scientist who did some of the original experiments that established the existence of radio signals. He also helped determine the important characteristics of these signals. These old-timers gave us radio technology as we know it today and are honored by having many of the units of measurement named after them. For some interesting reading, type in a name such as Tesla, Hertz, or Marconi at the Wikipedia Web site or with a search engine such as Google.

Radio waves have frequencies that seem incredibly high compared to the kind of waves you may be familiar with, such as sound waves. Humans can hear sound waves with frequencies of about 20 Hz (20 cycles per second) to 20 *kilohertz* or kHz (20,000 cycles per second) — see the sidebar about the metric system. Radio signals, on the other hand, generally start at a 20 kHz frequency, and they continue up through the thousands, millions, and billions of hertz — and beyond!

In fact, visible light is a kind of radio wave that has a frequency about a million times higher than radio waves. It boggles the mind just to think about it! The radios I talk about in this book use frequencies between 100 kHz (500,000 cycles per second) and 1 *gigahertz,* or GHz (1 billion cycles per second). A signal in this range of frequencies is referred to as *radio frequency* or *RF signal*, *RF* for short.

The huge swath of frequencies used by radios is known as the *radio spectrum*. The spectrum is too big to consider in one chunk, so radio users have divvied it up into ranges called *bands* in which the signals have similar properties. Table 2-1 shows the various bands. When I talk about the different kinds of radios, you can see how these ranges relate to how the radios perform.

Table 2-1	Bands in the Radio Spectrum	
Frequency Band	*Range of Frequencies*	*Typical Use*
Medium frequency or MF	300 kHz to 3 MHz	AM broadcasts
High frequency or HF (the traditional *shortwaves*)	3 MHz to 30 MHz	Shortwave broadcasts, such as the BBC
Very high frequency or VHF	30 MHz to 300 MHz	FM radio and VHF TV channels
Ultra high frequency or UHF	300 MHz to 3 GHz	UHF TV and mobile phones
Microwave	Above 3 GHz	Satellite TV

In the U.S., the FCC is in charge of administering access to all these frequencies, and literally hundreds of different groups all have their own little homes on the range, called *allocations*. When you begin using your radio, you'll be in one of those groups, too. Scanner fans and shortwave listeners need to know where their quarry can be found on the radio dial. I show you specific allocations for each type of radio in Parts II, III, and IV. In the meantime, if you'd like to take a look at an amazing chart showing how the spectrum is divided up, browse to `www.ntia.doc.gov/osmhome/allochrt.pdf` and see if you can find the bands where the AM, FM, and TV broadcasters transmit!

The metric system

In this book, and everywhere that radio technology and radios are described, the metric system is the standard way of describing numbers and measurements in *metric units*, such as the meter, which is abbreviated *m* and is a little more than 3 feet (or 1 yard) long. The metric system also uses a set of *prefixes* that make the unit larger or smaller. Metric prefixes are very convenient for handling numbers from the very small to the very large. The following list gives you the most common metric prefixes and abbreviations.

Ohms are a measure of electrical resistance and watts measure power. Amps and volts measure electricity. Pico and nano are measurements used with electronic components.

Prefix and Meaning

k, kilo (1,000)

M, mega (1,000,000)

G, giga (1,000,000,000)

c, centi (.01, or one-hundredth)

m, milli (.001, or one-thousandth)

μ, micro (.000001, or one-millionth)

n, nano (.000000001, or one-billionth)

p, pico (one trillionth)

Common Abbreviations

km, kHz, kohms or kΩ, kW (kilowatts)

MHz, M ohms or MΩ

GHz

cm

mm, mamps or mA, mV (millivolts), mW (milliwatts)

μA (microamps), μV (microvolts), μW (microwatts)

Understanding wavelengths

Earlier in this chapter, I describe a radio wave as a *traveling vibration.* Not only do radio waves vibrate at high frequencies, but they move fast, too. They move at the speed of light because (wrap your head around this one!) radio waves are a low-frequency form of light! How fast is the speed of light? 186,200 miles-per-second — a rather ungainly number. The metric system makes it a little easier to remember how fast light moves — 300 million meters per second or 300 meters per microsecond. This is so fast that a radio wave could circle the earth in ⅐ of a second or travel to the moon and back in about 2½ seconds. All radio waves travel at this same speed, and all radio waves move a little slower when traveling along wires and cables.

In "Introducing frequency," earlier in this chapter, I suggest that you tie a string to a stationary object and pull the string so that it is not loose (but don't make it too taut, either). Well, you should head back to your string for another experiment if you want to understand a little more about wavelengths. When you pluck the string, take a close look at that wave as it travels away from you. As long as you keep the string at the same tightness, the length of the ripples stay the same, whether you pluck softly or strongly, just as a guitar string makes the same note whether loud or soft. The length of the ripple as the string makes one complete vibration is called the *wavelength*.

Because radio waves all move at the same speed (that of light), wavelength is just another way of talking about how fast the wave (or the string, if you're playing around) is vibrating. If the frequency is low, the wave takes longer to make one cycle and that means the radio wave travels for a longer distance while doing so. If the frequency is high, the cycle is completed quickly and the wave only travels a short distance. Figure 2-1 shows how this relationship works. Radio waves have wavelengths between 600 meters (for a 500-kHz signal) and 3 cm (0.03 meter, for a 1-GHz signal).

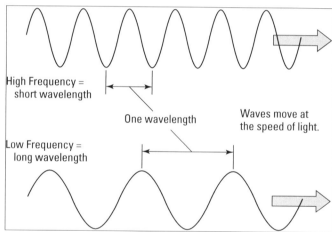

Figure 2-1:
The higher the frequency, the shorter the wavelength, and vice versa.

To convert frequency measurements to wavelength use one of these simple methods:

✔ To convert from MHz to meters, divide 300 by the frequency. For a 100-MHz signal, this is your calculation:

```
300 ÷ 100 = 3 meters
```

✔ To convert from meters to MHz, divide 300 by the wavelength. For a 10-meter signal, this is your calculation:

```
300 ÷ 10 = 30 MHz
```

Any radio signal can be described either by its frequency (which is most common) or by its wavelength. In this book, I usually use frequency, only changing to wavelength when physical distance or length is important, such as when discussing antennas.

Getting From Here to There: Propagation

The movement of radio signals from point to point, whether near or far, is known as *propagation*. Because the propagation varies dramatically with frequency, the different types of radio waves have very different characteristics. Some signals can propagate worldwide, while others are limited to just a few miles at most.

Signals on the ground and in the sky

Close to an antenna, the waves of a radio signal spread out at the speed of light like ripples in a pond. The radio signal moves along the ground and into the air. Signals hugging the ground, called *ground wave signals*, travel until the soil and plants absorb them. If you have an AM radio receiver, you receive AM broadcasts during the day from signals sent along the ground out to about 100 miles from the station.

As frequency increases, ground wave signals die away much quicker until, at the lower ranges of the VHF bands, the ground wave signals become useless. Check out Table 2-2.

In the VHF, UHF, and microwave bands (see "Measuring frequency across the spectrum" for a reminder of what frequencies make up the bands), the atmosphere acts as an absorber. The higher the frequency, the more the signals are absorbed by water in the form of rain, fog, or just vapor. Here's how these signals propagate:

The signals spread out in straight lines away from the source, traveling higher in the atmosphere as the earth curves away below them.

Eventually, these signals leave the atmosphere and what is left of them speeds off into space, carrying your conversation as well as episodes of *Baywatch* and the nightly news to the stars.

For that reason, propagation on these bands is considered to be primarily *line-of-sight*. Beyond the horizon, signals are lost to the earth's curvature.

Table 2-2	Propagation Patterns According to Frequency	
Frequency	*Signal Propagation*	*Distance Signals Can Travel*
VHF, UHF, and Microwave	Signals are absorbed by atmosphere or are lost beyond the horizon as the earth's surfaces curves.	VHF can travel 100 miles; UHF can travel 20 miles; Microwave can travel 5 to 10 miles.
MF and HF	Signals travel along the ground or bounce off the radio mirror of the ionosphere.	Ground waves can travel 5 miles or hundreds of miles depending on conditions. Sky waves can travel worldwide at times.

Below the VHF bands, a different effect takes over, bringing some surprising results:

✔ Absorption in the atmosphere is much smaller, so the signals retain more of their strength as they travel.

✔ The upper atmosphere is exposed to the strong ultraviolet radiation from the sun, which knocks electrons (remember them?) free from their host atoms. There are enough of these free electrons that they act like an electrical mirror to radio waves if the frequency is low enough.

This region, from 50 to 250 miles above the earth, is called the *ionosphere*. Signals in the HF bands and at lower frequencies can be reflected back to earth when they encounter the ionosphere, returning to the ground hundreds of miles away or more! This *sky-wave* propagation is often called *skip* and allows long-distance communication by shortwave stations, ham radio operators, Citizens Band enthusiasts and others. When you hear a weak and faraway AM station coming through your radio at night, you're listening to a sky-wave signal! (For more information about long-distance propagation, see Chapter 15.)

Reflections

If you read the previous section about how low-frequency signals can be reflected by a cloud of electrons, you may wonder whether these signals can also bounce off of other things. You bet! Signals are reflected by buildings, mountains, the ocean, water towers, jetliners, and even the moon — anything that looks different to a radio wave than whatever it has been traveling through. Even plain old dirt reflects radio waves, although it absorbs some of them.

Reflections are particularly strong at frequencies with short wavelengths, such as those at and above the VHF bands.

Reflections may help your signal get from place to place when the direct path between you and the distant station is blocked. For example, I've bounced signals off Mount Rainier to get a signal from Seattle, Washington, to Portland, Oregon!

Of course, there's always a downside. When reflections create more than one path between the two points, the different signals can interfere with each other, reducing or canceling the received signal! This is called *multipath interference*. You can observe this interference yourself by dripping water into a large bowl of water and watching the resulting pattern in the waves as they bounce outward from the droplet and bounce off the sides of the bowl. At some points, the reflected waves are both moving up or down at the same time, so they add, and in other places they cancel out, just like radio waves do.

The weather, the sun, and the seasons

The atmosphere tends to absorb radio signals at high frequencies (the higher the frequency, in fact, the more quickly the atmosphere absorbs the signals). However, any kind of a change encountered by a radio wave tends to cause some of the signal to be reflected. As you know, the atmosphere isn't a completely uniform bunch of gas; it has clouds, storms, hot and cold spots — weather. All those elements create a *dynamic* (rather than *static*) atmosphere that is constantly changing, with no two areas exactly the same. Here are some ways in which changes in the atmosphere can affect radio signals:

- **Weather:** When a radio wave encounters a change, some of the signal gets reflected. If the radio wave hits the region of change just right, sometimes it will travel along the change for a while, like a skateboarder riding a stair rail, before breaking loose somewhere faraway. Radio mavens know that this happens most frequently for VHF and UHF signals, so they watch the weather for the right conditions and then try to listen or communicate over long distances otherwise unspannable. Because this sort of propagation takes place in the part of the atmosphere known as the *troposphere*, it is called *tropospheric propagation*.

- **Ultraviolet light:** The sun's ultraviolet radiation is what creates the ionosphere, which reflects HF signals back to earth. That means that there must be strong day and night variations in propagation. You can almost hear the planet turning as you listen to shortwave signals from faraway. As the sun illuminates different parts of the globe, a signal fades or grows, sometimes strong for hours and other times for just minutes.

The sun's 11-year *sunspot cycle* also strongly affects the ionosphere's ability to reflect signals. At the cycle's peak (one just occurred in 2002), there was so much ultraviolet radiation hitting the earth that signals on some frequencies could be heard all over the world 24 hours a day! As we reach the bottom of the cycle (estimated to come around in late 2006), those previously strong signals won't be heard at all.

✔ **Seasons:** In addition to the daily variations of daylight and weather and the years-long cycle of the sunspots, radio waves are also affected by seasonal variations. Because the earth's axis is tilted, the amount of sunlight a particular geographic region receives varies between summer and winter. This variation affects the strength of the ionosphere, and thus, its reflecting qualities. Ham radio operators, in particular, know this variation well, scheduling their activities and competitions to take advantage of the best conditions.

✔ **Meteors:** Seasonal variations also extend to the stars themselves in the form of meteors! When a meteoroid burns up in the atmosphere, becoming a meteor, it leaves a trail of super-hot gas that can reflect radio signals for a short time (up to ten seconds or so). When this happens, two stations lying in wait at ground level can exchange a short burst of data. With thousands of meteors traveling through the atmosphere every day (most quite small), stations can take advantage of meteors more often than you may think. The meteor showers that occur at well-known dates throughout the year are prime periods for *meteor scatter* communication.

This list just touches the surface of the different ways that radio signals travel from one radio to another! The science of propagation touches every part of what we know about the sun, space, and the makeup of the earth.

What You Hear Is What You Get: Modulation

If you look at the tuning controls of a shortwave radio or in the broadcast station listings of a newspaper, you'll see the notations *AM* and *FM*. These mean *amplitude modulation* and *frequency modulation*, respectively. By itself, a radio signal doesn't carry any information, such as voices or pictures. The information has to be added to the radio wave somehow. *Modulation* is the process of adding that information and *demodulation* the process by which the information is recovered. The following sections explain these concepts in further detail.

Amplitude modulation (AM)

The oldest and simplest way of adding information to a radio wave is to make the wave stronger or weaker in a way that represents the information you're trying to communicate. The strength of a radio wave is called its *amplitude,* so adding information to the wave by varying its strength is called *amplitude modulation,* or *AM* for short.

Say the word *hullaballoo* out loud and note the pattern of loud and soft sounds, as well as the different frequencies in your voice. If you were speaking into a microphone for broadcast to an AM radio, by *modulating* the radio wave's amplitude with these various frequencies and the loud and soft patterns of your voice, the information you're conveying (the word *hullabaloo*) is added to the radio wave and transmitted far and wide! Figure 2-2 illustrates AM.

If you watch a meter that measures transmitter power while you speak into the microphone of an AM transmitter, you'll see the meter's needle jump up and down as your voice gets louder and softer, just like on a tape recorder. Because broadcast stations using AM are grouped together between 550 and 1700 kHz, this is referred to as the *AM band.*

You may notice when you listen to AM radio (say while driving in your car) that there is a bit of interference, as well as static crashes. This is because the various noises from natural sources, such as lightning, and from man-made sources, such as you car's spark plugs, are amplitude variations, just like signals. Your radio can't tell the difference between the two.

Morse code, the ancient modulation

The simplest way to change the amplitude of a radio signal is to just turn it on and off. That's what Morse code does. By turning the signal on and off in a pattern that represents individual characters, a text message can be sent over the air. Sending signals in such a way is what the pioneers of radio did at the beginning of the 20th century, adapting telegraphy to the new communications medium of radio.

In fact, the original term *wireless* comes from wireless telegraphy. Morse code is alive and well today, used by many ham radio operators around the world. Also called *CW* for *continuous wave* (another early radio term), Morse code transmissions can be generated with extremely simple transmitters. Because it is so easy to detect and listen to, Morse spans a greater distance with lower power and cost than any other type of modulation. I use it all the time! For some fascinating reading about the history of radio, browse to `inventors.about.com/library/inventors/blradio.htm`. A good book about the development of radio and broadcasting is *The Early History of Radio: from Faraday to Marconi,* by G. R. M. Garratt (Institution of Electrical Engineers). My book, *Ham Radio For Dummies* includes a lot of information about how CW is used today.

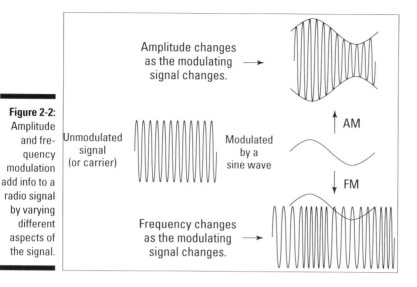

Figure 2-2:
Amplitude and fre-quency modulation add info to a radio signal by varying different aspects of the signal.

Frequency modulation (FM)

Frequency modulation, or *FM*, adds desired information to a signal by varying the signal's frequency. In fact, the amplitude of an FM signal doesn't change at all, no matter how loud or soft the modulating information may be. The only thing that changes is the frequency of the signal.

An FM receiver can just ignore AM signals, because it only cares about variations in frequency. This makes the recovered signals almost noise free! This is one of the advantages of FM, and a reason for its popularity. (FM is the most widely used modulation of all, despite the complexity of FM radio equipment.)

Like AM broadcast stations, FM stations are grouped together in their own band — 88 to 108 MHz, just between TV channel 6 and the aircraft band. This range of frequencies is called the *FM band.* An AM-FM radio receives AM in the lower range of frequencies and FM in the higher range.

Terms of Endearment: Using the Language of Radio

Although a complete glossary is provided in the Appendix, the definitions in this section can help you understand the basics of operating your radio. All these terms are common to many radios. Of course, I go into more depth in my explanations in the chapters covering each type of radio.

Understanding controls and features

After you take your radio out of the box, all those controls have new and unfamiliar labels. Here are the ones you'll encounter first:

✔ **Gain:** A general term applied to all adjustable signal levels. *AF Gain* stands for audio-frequency gain and is the same as a volume control. *RF Gain* controls the sensitivity of the receiver. *Mic Gain* controls how much audio from the microphone is applied to the transmitter circuits.

✔ **Squelch and Busy:** *Squelch* is a control for the circuit that cuts off the speaker or headphone audio when no signal is present. This control keeps you from having to listen to the hiss and crackle of noise while you're waiting for a signal (over time, you'll come to appreciate this a lot more). When a signal is stronger than the squelch threshold, it *opens the squelch* and when the signal goes away the squelch *closes.* The squelch control adjusts the threshold at which the squelch opens. Raising the threshold *tightens the squelch* and lowering the threshold *loosens the squelch.* A *busy* indicator may light up to indicate when the squelch is open.

✔ **Push to Talk (PTT):** To cause a two-way radio to transmit, you must close a switch to activate the transmitter. The switch is usually a button or lever on the side of a handheld radio, although it may be a separate microphone. It must be closed (or pressed) when you transmit and released when you are listening.

✔ **Simplex, Duplex, and Repeater:** Terms that indicate forms of communication. If you are communicating directly with another radio user on a single frequency by taking turns talking and listening, that is *simplex* communications, meaning *one frequency at a time. Duplex* communications takes place on two frequencies, one for transmitting, and one for receiving. This form of communications is used when communicating with the assistance of a *repeater,* which relays your transmitted signal over a wide area.

The antenna

You can't have radio without antennas! Here are definitions for a few of the most common antenna terms:

✔ **Vertical or Ground Plane:** An antenna that is oriented perpendicular to the earth's surface. The term *ground plane* is used when the antenna requires a conducting surface underneath it as part of its normal operation. Vertical antennas are usually *omnidirectional,* meaning that they transmit and receive equally well in all directions broadside to the antenna.

- **Beam:** An antenna that focuses its transmitted energy and receives best in one direction. Beams are *directional* antennas.

- **Rubber duck:** Slang for the plastic-covered flexible antennas on hand-held radios.

- **Whip:** A slender metal rod antenna, stiff enough to maintain a straight shape without any additional support. Some are retractable (think *car antenna*) and others are not.

- **Mag mount:** A vertical whip antenna attached to a magnetic base that is held onto a metal surface, such as a car's roof or trunk.

The contact

The more things change, the more they stay the same. Here is some very old radio jargon that is a key part of the radio etiquette that keeps things flowing smoothly over the airwaves:

- **Call and Sign:** The set of letters and numbers on your FCC license is your *call sign*, or *call* for short. When you attempt to make contact with another station by transmitting your call sign, you are *calling* or *making a call*. When you give your call sign in order to identify your station, you're *signing*. When you give your call sign before concluding a contact or leaving the air, you're *signing off*.

- **Over, Clear, and Out:** Because two-way radios can only receive or transmit (but not both at the same time), you have to give the other station a cue that you're done speaking and not just pausing. When you are finished speaking, say, "Over." If you are concluding your contact and making the frequency available, but you're not leaving the air, say, "Clear," to let others know that you will still be listening. If you are leaving the air entirely, say, "Out." It's not necessary to say, "Over and clear," or, "Over and out." Those aren't good radio etiquette.

- **Roger:** Technically, this term only means "I received and understood your transmission." It does not mean "Yes," or "I agree," or "I will respond."

Chapter 3

Making Radio Fit Your Life

*S*ervice is the name the Federal Communications Commission (FCC) uses for the package of frequencies, signal types, operating rules, and licensing regulations that apply to a specific class of radio communications. Grouping radio users based on the various services they use is necessary to prevent chaos on the airwaves. Imagine a broadcaster, a delivery truck, and volunteers at a parade all trying to use the same frequency! By grouping similar communications needs together, everybody gets much better results with less interference and simpler rules. There are many services and they are subdivided further into groups of similar users, each with its own chunk of radio frequencies, called an *allocation*. Each service and allocation has its own set of rules, some of which apply to the operator, as well!

Radios are made to be used in a specific allocation and have features that follow or *comply* with its rules. For example, the handheld radios for use in the *Family Radio Service,* or *FRS,* can only operate within a set of a few frequencies (called *channels*) in the ultra high frequency (UHF) band. These radios must comply with certain rules. For example, there are specific limits on how much power they can generate, what antennas they can use, and other technical aspects.

If you already have a radio, understanding the service and the allocation it's made for helps you make the best use of the radio. If you're looking for a radio to fill a specific need (even if the need is simply the need to tinker and experiment with radio), you need to understand the different services and allocations in order to select the proper one. This chapter explains the categories, introduces you to rules, regulations, and licensing requirements, and helps you decide on the service that best suits your needs.

Seeing What Makes Radio Services and Allocations Different

Each service has unique characteristics that set it apart. The different sets of characteristics were chosen in order to satisfy the communications needs of each specific group of radio users. For example, broadcasters are set apart from other users so that the public can easily find them. Each type — AM, FM, TV — is assigned an allocation that is well suited to serving the desired audience. For example, taxis and police radios are required to operate on the very high frequency (VHF) and ultra high frequency (UHF) bands, using small antennas so that their signals stay within a city or town. After all, Floridian taxi drivers do not need to hear Chicago dispatchers.

If you didn't take a look at the FCC's chart of frequency allocations while reading Chapter 2, this would be a great time to have a look — the chart put out by the U.S. Department of Commerce located at `www.ntia.doc.gov/osmhome/allochrt.html` gives you a glimpse at the complex system of services and the allocations each receives.

Radio services fit broad categories such as Mobile, Fixed, Amateur, Broadcasting, and Maritime Mobile. Within these categories lies a cornucopia of allocations. For example, the HF CB service (that's high frequency Citizens Band) is allocated a particular set of frequencies as part of the fixed and mobile services.

Characterizing services by frequency and modulation

A single service, such as the Land Mobile Service, may have several allocations that support different types of communication needs. The primary characteristic that defines an allocation is the needs of the users. When you think about it, it makes sense. Certain types of radio communications will simply work better at certain frequencies, so these frequencies ought to be dedicated to those uses. Each group of users gets access to a band of frequencies that may be *shared* with users from other services.

You can find the high frequency (HF) Citizens Band (CB) service shown as a pair of pink slices labeled Mobile near the right side of the third strip from the top of the U.S. Department of Commerce chart (yes, Virginia, the chart really is as complicated as all that). Figure 3-1 gives an example of how the allocations are labeled. The HF CB frequency allocation is from 26.96 to

27.41 MHz. You can also see that the upper half of that range (from 27.23 to 27.41 MHz) is also shared with stations in the Fixed radio service. (The service names on this chart are broad categories, so you won't find the individual allocations listed here explicitly. Darker colors in the far left corner indicate government services.)

Along with restrictions on frequency, the different services are also restricted by the type of modulation their radios can employ. CB radios can only use AM or a more refined type of AM called *single sideband (SSB)*. When you review the frequency allocations provided by the U.S. Department of Commerce, you see that just below and to the left of CB are two big, blue blocks that represent two broadcasting service frequencies that are right next to each other. The one from 76 to 88 MHz is for TV Channels 5 and 6, while the one from 88 to 108 MHz is for FM broadcasters. These are both broadcasters, but they must use very different types of transmitters.

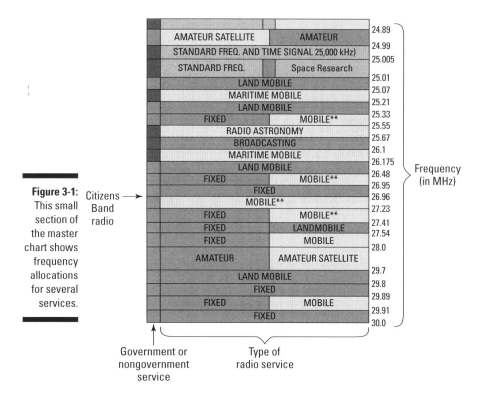

Figure 3-1: This small section of the master chart shows frequency allocations for several services.

Licensing requirements and operating rules

Some of the services, such as the Marine, Amateur, and General Mobile Radio, require that users obtain a license to operate a station. By administering a license program, the FCC is able to manage the demand for radio communications and ensure that the airwaves are used appropriately. You have to pay a fee for your license, as listed in the section, "Getting a License When You Need To."

Unlicensed services, such as Citizens Band, provide a means for you to communicate, but don't guarantee any particular quality. It's important that you play by the licensing rules, only in part to keep order on the airwaves. Unlicensed operation can result in the loss of your equipment or significant fines.

Whether or not the service is licensed, the FCC has also created operating rules for some of the services. These specify what sorts of communication may take place over the airwaves, how the stations identify themselves, who they may communicate with, and so on. I note significant rules for each service I mention in this chapter so that you know what you're in for before you select a service for your needs.

Characterizing services based on equipment limitations

Each service is characterized by several rules in addition to frequency and modulation (see the previous section). These rules pertain to the equipment that may be used by each service in its allocations. Here's the rundown on equipment that's regulated:

- Output power.
- Antennas.
- The exact frequency and spacing of the individual channels on which the radio operates.

  For example, Family Radio Service radios may output no more than a ½ watt of power and the antenna must be permanently attached to the radio. This restriction makes the radios useful for short-range communications, but prevents interference to other users in the region.

The radios must also be *type accepted*, which means that their design and construction must be acceptable to the FCC. Only certified technicians are allowed to modify or repair type-accepted radios. The reason for this rule is not to make certified radio technicians rich; it is intended to guarantee that users will not cause interference from faulty or *noncompliant* equipment that does not meet the technical standards.

Comparing Two-Way Radio Services

The following list gives you a sense of the range of services available. I discuss these services in more detail in Part II, but this list helps you get a feel for what each is like. You can get a feel for each service's primary characteristics when you compare it to Table 3-1.

- **Family Radio Service (FRS):** Used for unlicensed, short-range personal communications by low-power two-way radios.

- **General Mobile Radio Service (GMRS):** Similar to the FRS, but requires a license and allows more power and better antennas.

- **Multi-Use Radio Service (MURS):** An unlicensed service for short-range business and personal communications.

- **Business Radio Service (BRS):** Full-featured business and public safety use for mobile and fixed-station communications.

- **Citizens Band (CB):** An unlicensed service for personal and business needs over a few miles range.

- **Marine VHF:** Licensed and unlicensed use by recreational and commercial mariners for local and regional areas.

- **Marine HF:** Long-distance, licensed service for mariners.

- **Amateur Radio:** Licensed service for individuals to experiment, provide emergency communication, and become trained in radio communications.

Table 3-1			Primary Characteristics of Common Radio Services			
Service	Range	Frequency	Business Use?	License Required?	Data Transfer Okay?	Comment
FRS	<½ mile	UHF	No	No	No	Short-range, personal only
GMRS	Up to several miles	UHF	Yes	Yes, $80	Yes	Repeaters increase range
CB	<10 miles	HF	Yes	No	No	Skip propagation can cover long distances
MURS	Up to several miles	VHF	Yes	No	Yes	Business and personal
PLMR	Up to regional coverage	VHF, UHF	Yes	Yes, $55 plus coordination fees	No	
Amateur Radio	Up to world wide	HF, VHF, UHF	No	Yes, $14 (10 year)	Yes	
Marine HF	Up to world- wide	HF	Yes	Yes, $205 for the ship, $55 for the operator license	Yes, some channels	Long-distance and regional
Marine VHF	<25 miles	VHF	Mixed, $55 if required (see FCC rules)	Optional	No	

Choosing Between the Services

If you bought this book, chances are you have *some* idea what you would like to do. But if you're just exploring the radio scene (oh, yes, there is a scene!) and can't decide what service or services you're most interested in, start by deciding what you are trying to accomplish with the radios. Perhaps you just want to have fun. Or maybe a radio would help out a lot at work. In either case, knowing your necessary *range* (how far your signals travel) is essential. Ask yourself these questions about range:

- ✔ Assuming you want two-way communications, are you trying to keep in touch with family members close by or around town?
- ✔ Is the radio for emergency communications in your neighborhood?

Here are some considerations based on your potential interest:

- ✔ **Listen-only users:** Listen-only radio enthusiasts need to consider service-related issues. Maybe you want to receive the local public safety transmissions and weather alerts that are broadcast using FM signals. No problem. But if you want to add in the aviation channels, the receiver has to handle AM signals, too. The services are spread out over a very wide frequency range, so be sure you get a receiver that can tune everywhere you need it to!
- ✔ **Business two-way radio users:** If you plan on using the radios for your or an employer's business, make sure that you pick a service that allows business-related communications. In some nonbusiness services, such as ham radio, it's strictly forbidden to use the radio to conduct business activities, although it's okay to say, "I had a tough day at work!" over the air.

Buying Equipment to Fit Your Budget

Buying new equipment is the safest option, because you'll get the manufacturer's warranty and probably some support from the dealer. I always recommend buying new equipment to beginners because beginners haven't yet attained the level of experience necessary to test or troubleshoot old, cranky equipment. If you can buddy up with a knowledgeable friend, there are some great bargains out there, however.

Leave room in your budget for accessories and antennas. Batteries and power supplies can cost as much as half of the radio. Antennas and connecting cable range from affordable to quite expensive, and become particularly expensive if you have to mount them outside on a permanent structure of some kind. Basic accessories and installation materials will cost from one-third to one-half of the radio's price (or up to $200). As you get into the more expensive radios, the fraction will drop, but the cost will still be substantial.

Your operating needs will also affect your radio equipment choices. If you want to operate or listen while away from home, be sure to acquire a radio that is designed for that kind of use. A radio intended for fixed-location use that requires 115 V ac power may be difficult to use while traveling.

Getting a License When You Need To

What's the point of using a service that requires you to get licensed? Consider the difference between a commercial truck driver permit and a private driver's license. The first is highly regulated and performance standards are strictly enforced. On the other hand, nearly anyone can get a driver's license, so performance and safety are often erratic. It's a trade-off of convenience against performance. Licensed services are held to higher standards, with a better quality of communications, than unlicensed services.

Users of licensed services are also protected against interference from users of unlicensed services, by law. Unlicensed services have no such protection. Users of unlicensed services must not cause interference and must accept any interference they may receive (in other words, deal with it!).

This is true for *all* unlicensed radios, including appliances such as cordless phones and baby monitors. Don't believe me? Dig out your cordless phone's owner's manual and look for a paragraph about it. You may even see a sticker on the appliance itself.

If you need high-quality communications, consider a licensed service. If you are using or planning to use an unlicensed service or will just be listening, you can skip the rest of this chapter.

Who made them king? — The FCC

The Federal Communications Commission was created by the Communications Act of 1934 and given ultimate authority over communications in the United States — *all* communications. So there! There is no higher communications authority in the U.S. Internationally, each country has its own similar government agency, often called the PTT, which stands for *Post, Telephone, and Telegraph*.

The FCC *is* listening

Unlicensed operation is an area in which the FCC Enforcement Bureau is increasingly active. In other words, the Enforcement Bureau looks for people who should be licensed but aren't. Think twice before deciding you really don't need to pay that license fee.

On a related, but useful, note, imported electronics do not always comply with FCC rules (although they are supposed to) and frequency allocations. Before purchasing that ultra-bargain wireless gadget, be sure it operates according to FCC rules so you don't accidentally cause interference and attract the FCC's attention! Is the FCC really listening? Ask some of the trucking firms whose drivers were caught using high-powered ham radios without a license rather than legitimate business or CB radios! Fines starting at $10,000 caused those illegal radios to get yanked pretty quickly.

The FCC has the authority not only to make regulations, but also to enforce them, and is allowed to levy and collect fines from miscreants. Local and state regulatory agencies cannot contradict FCC requirements or prevent radio operations, although local land-use and zoning laws can place constraints on radio operation. The FCC also has the responsibility to ensure that its regulations are reasonable and foster a healthy communications environment. All in all, the FCC does a pretty good job, even though its resources are stretched thin in places.

Registering with the FCC online

Except for the Amateur Radio Service, you don't have to take an exam to get an FCC license. Rather, all you need do is acquire the proper form for the selected service, fill it out, and send a check. Soon, your licensing paperwork appears in the mail. Actually, you can begin transmitting as soon as you see your call sign in the FCC's master database.

You must register yourself online with the FCC and obtain a *Federal Registration Number* or *FRN* before you can do anything else. The FRN is the master identification number that associates you with a specific license. Your FRN can be associated with more than one license, if necessary. You need an FRN in order to use the online applications for particular radio services. Your FRN is not private, although the password you select is.

To obtain your FRN, visit the following site and follow these steps after you connect to `svartifoss2.fcc.gov/servlet/coresweb.servlet.user.HomeServlet`.

1. **Click the Register button.**

   You're asked two questions. First, you're asked whether you want the license for business use or for personal use. The other question you're asked is whether your contact address is within the U.S. or its territories.

2. **Select the appropriate radio buttons to answer these questions.**

3. **Click Continue.**

4. **Fill in the appropriate information on the page that appears on-screen.**

   If you are registering a business, you have to include information about the business. Otherwise, if you're registering as an individual, you only have to fill in basic information, including your name, address, and Social Security Number.

   At the bottom of this page, you're given an option to come up with your own, unique, case-sensitive password (made up of 6 to 15 characters) and a password hint. Or you can have the FCC come up with a password and hint for you (skip to Step 6).

5. **If you want to come up with your own password, type it and then reenter it in the appropriate text boxes. Type your hint.**

6. **Click the Submit button.**

   If you have chosen to allow the FCC to choose a password for you, a dialog box appears on-screen to ask whether you're sure that you want to have the FCC do this.

7. **Click Yes if you're sure. Otherwise, click No and go back and write your own password.**

   A screen appears with your FRN. You're now ready to register for a particular FCC radio service.

Accessing the ULS and applying for a new license

You can access the FCC license database via the FCC's online *Universal Licensing System,* or *ULS.* The ULS home page is shown in Figure 3-2. From this page, you can search through the license database, apply for a license for a particular service, and pay any necessary fees.

You can't apply for a license for a particular service at the ULS without first registering with the FCC and obtaining your FCC registration number (or FRN). See "Registering with the FCC online" for more information.

Figure 3-2:
The home
page of the
FCC's
Universal
Licensing
System.

If you choose not to acquire an FRN (why would you?), you'll have to do business the old-fashioned way — on paper.

After you've acquired an FRN, apply for a license by following these steps:

1. **Visit wireless.fcc.gov/uls.**

2. **Click the On-Line Filing link on the ULS home page.**

   You are directed to the License Manager system.

3. **Enter your FRN and password to log on to the License Manager.**

   If you forgot your FRN number or password, you can click the appropriate link and have your hint sent to you.

   If you have no current licenses associated with your FRN, the My Licenses page displays a message to that effect. You're offered two options — one to associate existing licenses with the FRN, and the other to obtain a new license.

4. **Click the Apply for a New License link.**

   The Apply for a New License page appears, along with a drop-down list of all the license options available.

5. **Choose one of the services from the Select Service drop-down list, and then click Continue.**

   If, for some reason, you need to go back, you can click the Back button.

   Depending on which service you've selected, you'll be guided through a series of questions and data entry pages. On the right side of the browser window, a Steps window shows you where you are in the licensing process. A Questions window on the lower-right side of the screen answers questions about the applications process.

   For example, if you are applying for a GMRS license, you're prompted to provide your name and address and certify that you will follow the rules.

6. **When you've completed the application, certify that all the information you've provided is correct.**

   You're handed off to the fee payment system (Form 159), where you can use a credit card in a secure payment system.

If you get stuck or confused at any point, I suggest that you ask the dealer where you bought your radio for help. If you are applying for a marine radio license, the folks at marine electronics shops are usually quite knowledgeable and can help you file the necessary forms.

When you're finishing paying your fee, a few days may pass before the FCC actually generates your call sign. You can check to see whether you have a call sign by returning to the FCC ULS home page and using the Search function, entering your name and hometown. The page displays a message that says `application pending` until the license has been granted. After that, your call sign appears. You're on the air! A paper copy of your license will arrive via snail mail. If, for some reason, the FCC encounters problems during the registration process, you will be notified by e-mail and snail-mail to return to the application process and fix specific problems.

Call signs in the United States begin with the letters N, K, or W (amateur radio users — *hams* — also get call signs beginning with A) and may have up to three letters followed by three to five numbers. For example, WGN4321 is a call sign. Only broadcast stations get letter-only calls and only hams are allowed to choose their own call signs.

Coloring Inside the Lines: Basic Rules

All the radio services that an individual or small business is likely to use have the same basic ground rules. If you play by these simple rules, you'll have little to worry about. There are really very few operating rules and they're easy to follow.

Broadcasting (one-way transmissions)

Everybody wants to know, "Can I play music or host my own talk show on my radio?" In general, the answer is no. By continuously transmitting, you tie up a precious radio channel, denying its use to others who may have more urgent needs to use it. There are broadcasting services that are more appropriate outlets for your creative urges.

Identifying your station

If you are using a licensed service, you need to know about a few specific rules regarding identifying your station with the FCC-assigned call sign.

Most of the rules regarding identifications are straightforward. For example, you're invariably asked to say "This is [your call sign]" whenever you begin and end a conversation. If a conversation continues for a long time, you will need to identify your station at regular intervals. Not only is this required, it's good practice on a shared channel so that others can tell who is transmitting.

Recognizing power and antenna limits

Each service has specific limits on transmitter output power. Don't use *bootleg amplifiers* that increase the output power of your radio beyond the limits for your service and allocation — they usually have poor signal quality and cause a lot of interference to other radio users, not to mention the fact that they're illegal.

The service rules may also put limits on antenna height and type. These rules were created to allow many users to share the available channels. By limiting power and antenna height, each user is limited to a reasonable range.

Modifying your radio

Radios are type accepted to ensure that they work properly and meet the standards of the service. Unless you're a certified technician or experienced with electronics, don't tinker with your radio. You can find plenty of accessories and antennas to work with if you're a hands-on type.

If you *really* like to tinker, you're a natural for ham radio — so check out Chapter 9.

Avoiding naughty talk

If you wouldn't say it to your mom, it doesn't belong on the radio. Remember that you have no idea who might be listening to your conversations, so you should not give out personal or financial information, either. The airwaves are public and what you have to say can go a lot farther than you may think. Even broadcasters have to play by the rules that limit obscene and sexual content.

Where to Find All the Rules and Regulations

You can find most of what you need to know about the most important rules in your radio's operating manual or on the manufacturer's Web site. Remember that the rules are made to be relatively easy to follow. The FCC isn't trying to trip you up! If you received a set of license rules from the FCC with your license, keep that handy, too. Your radio dealer is also a good source of information.

Would you like to have a complete copy of the actual rules and regulations? Well then, browse right over to the FCC's ULS Web site `wireless.fcc.gov/uls` and download them to your heart's content!

On the left side of the ULS home page is a list of links. Click the <u>Radio Services</u>, and then click the link associated with the desired service. A page appears, filled with information for the service. A link to either licensing requirements or rules also appears on that page. For example, the GMRS page (`wireless.fcc.gov/services/personal/generalmobile/index.html`) has a link to the GMRS rules on the right side of the page, shown as <u>CFR, Part 95.1-95.181</u>. Click this link to be taken to a page that lists all the FCC rules. Click the <u>Part 95</u> link (for the GMRS) and a page with links to each rule appears. (These are also available in your radio's operating manual.)

The more popular services, such as CB radio, marine, and aircraft, all have a number of books written about them that go into detail about how to use those radios. I list a number of such references and Web sites throughout the book.

Part II
Two-Way Radios at Home, Work, and Play

The 5th Wave   By Rich Tennant

"Of course, your family radio system offers privacy codes and great clarity over a short range. But does it shoot silly string?"

In This Part . . .

Of course, there are quite a few different kinds of two-way radios out there, from low-power, short-range models you can hold in your hand to powerful units intended for worldwide communications. Part II gives you an under-the-hood look at each kind, beginning with the popular, handheld radios you see in department and sporting goods stores. Next up is Citizens Band, as useful as ever around town and along the roads. I also show you some important information about emergency communications.

Some businesses need radios too, so I cover the various types of radio systems that are aimed at those needs, including some new types that you may not have heard of before. Boaters are frequent radio users, relying on radios for safety and navigation, so there's a whole chapter on marine radio, both short range and long distance. Finally, I give you a peek into the amazing world of amateur radio, the oldest and most powerful radio service of all.

Chapter 4

A Radio in Your Pocket: FRS/GMRS Radios

*Y*ou can't go into a store these days without seeing a two pack of hand-held FRS/GMRS radios for sale. Walking around the mall, you may notice people using them in the same way they use mobile phones. Business staff use them instead of in-store intercoms, eliminating the annoying overhead PA system. You may also see them on the ski slopes, at the ballgame, or on the hiking trail. These radios have become quite popular, delivering effective, short-range communications at an affordable price.

This chapter addresses two major needs of FRS/GMRS radio users. First, it helps you decide on a radio (or get the best performance from the one you already own). Second, it cuts through all the licensing stuff in the operating manual, telling you just the facts you need to know in order to satisfy the FCC. I guarantee you'll have a new appreciation for your FRS or GMRS radio by the time you finish reading this chapter.

Introducing the FRS and GMRS Services

FRS stands for the Family Radio Service and GMRS stands for the General Mobile Radio Service. Now that we have *that* out of the way, what do these names mean and what are the differences between the two services? A *service* is a set of rules the FCC came up with when it created a specific means of communication. Each service was created with a certain communication purpose in mind, so the rules help the service function toward this purpose.

In the case of FRS and GMRS, the purpose is to allow individuals and businesses to use radio communications to conduct their affairs around home, store, or town. Table 4-1 summarizes the differences between the services. (In Europe, the PMR466 radios serve a very similar purpose although they're not precisely on the same frequencies.)

Table 4-1	An FRS/GMRS Comparison	
Capability	*Family Radio Service (FRS)*	*General Mobile Radio Service (GMRS)*
License required?	No	Yes ($80 for a five-year term)
How many channels?	14	22
Power	0.5 watt	5 watts (handheld); 50 watts (vehicle and base)
Range	Up to 1 mile under most conditions, more if in clear terrain or over water	Several miles with external antennas; wide area coverage with repeaters
Antennas	Permanently attached to radio	External base and mobile antennas allowed
Uses a repeater?	No	Yes (but repeater mode is not standard in all radio models)
Cost	$10 to $30 (usually sold in pairs)	$20 to $70 (combo FRS/GMRS); $150 (full-feature GMRS)

Radios that operate according to the FRS rules operate at a low power and are limited in their capabilities; they are most suitable for person-to-person communications over relatively short distances. GMRS radios are more powerful, so they have a longer range, and have more features that give them capabilities unavailable to FRS radios.

The most important technical differences between the two services are that GMRS radios may transmit with higher power, use repeaters (which I discuss later in this chapter), and attach mobile and base antennas. If you plan on making unlicensed radio communications an important part of your business or need to communicate reliably across distances of more than a mile, GMRS is the service you need.

These two services share a set of channels (fixed frequencies) between 460 and 470 MHz in the UHF range. The display on your radio shows the channel on which the radio is operating, from 1 to 14 on an FRS radio, and from 1 to 22 on a GMRS radio. The radios don't tell you what the actual frequency is — you have to look it up in your operating manual. Although you don't have to know the frequency to use the radio, it makes complying with the FCC rules easier if you do.

Here are the GMRS and FRS channel designation frequencies:

- Channel 1 = 462.5625
- Channel 2 = 462.5875
- Channel 3 = 462.6125
- Channel 4 = 462.6375
- Channel 5 = 462.6625
- Channel 6 = 462.6875
- Channel 7 = 462.7125

Here are the FRS-only channel designation frequencies:

- Channel 8 = 467.5625
- Channel 9 = 467.5875
- Channel 10 = 467.6125
- Channel 11 = 467.6375
- Channel 12 = 467.6625
- Channel 13 = 467.6875
- Channel 14 = 467.7125

To avoid interference, don't use a new radio until you've changed the channel setting. Most radios are programmed to remember the channel they were on when power was switched off. However, the first time you fire up a radio, it most likely defaults to Channel 1. If everyone uses the default channel, then no one's using any of the other channels. You'd be surprised at how many people just turn on a radio and start yakking, wondering why there are so many other people on the air!

Getting a GMRS License

In a nutshell, GMRS has a lot more going for it than FRS — more range, better antennas, and so on. The cost of GMRS radios isn't that much higher than it is for FRS radios, so the cost isn't prohibitive. What can be pesky is the license requirement (and the expense of obtaining the license).

You could just dial up a GMRS channel when you use an FRS/GMRS combo radio, right? Why not just buy the more capable GMRS units and start using them?

To be perfectly honest, the FCC is not driving unmarked vans around town looking for outlaw GMRS users. Its agents will, however, respond to complaints by licensed GMRS users about your unlicensed activity. It's questionable whether you would be assessed a fine (up to $10,000 for first offenses!) if you are caught bootlegging on GMRS frequencies, but comparing the potential hassle to a $75 license for five years (that's only $15 a year), licensing is the easiest (and most honest) way to go.

The Personal Radio Support Group (PRSG) at `www.provide.net/~prsg` runs a very good Web site that offers extensive discussions of FRS and GMRS rules. This site provides the complete rules for FRS and GMRS, as well as a FAQ page that goes into great detail about common questions.

Anyone can use FRS-only radios because they're limited to FRS channels and don't pack much power. Because combo radios that use both FRS and GMRS channels are available, you need to know the circumstances under which you should be licensed. Get a license if

✔ You plan to use any radio on a GMRS-only channel or channel pair.

✔ You plan to use a GMRS radio on an FRS channel while putting out more than ½ watt of power.

The operator's manual of your FRS/GMRS combo radio informs you that the radio automatically limits itself to the lower FRS power on FRS channels. Some allow you to increase the power on FRS channels if you have obtained a license.

Getting a license is not difficult. The entire process can be performed at the FCC's Universal Licensing Service (ULS) Web site, `wireless.fcc.gov/uls`. If you don't want to register online, you can register via snail-mail. The manual that comes with your radio also includes the procedure for getting a license.

To apply for your license

✔ **By paper:** Download the Form 605 in PDF format from `www.fcc.gov/formpage.html` or call 1-888-CALL-FCC. Fill out the form, attach your fee, and mail to the address on the form. In couple of weeks, your license is granted.

✔ **Online:** From the ULS home page, start by obtaining a Federal Registration Number or FRN, or FCC Registration Number, as described in Chapter 3. Your FRN is your identification number within the FCC

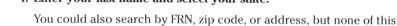

database and is associated with any FCC licenses you obtain. When you have your FRN, go back to the ULS home page and click the Online Filing button. If you already have an FRN, you can complete the GMRS registration procedure at `wireless.fcc.gov/services/personal/general mobile/licensing`. The steps are also outlined in Chapter 3.

After you apply, either by paper or online, you can periodically check the FCC database directly to see whether your license is in effect. Here's how:

1. **Visit the ULS home page (`wireless.fcc.gov/uls`).**

   Look for a Search heading.

2. **Click the Licenses button under the Search heading.**

   The License Search page appears.

3. **Click the <u>GMRS</u> link.**

   The GMRS License Search page appears.

4. **Enter your last name and select your state.**

   You could also search by FRN, zip code, or address, but none of this information is strictly necessary unless your name is John Smith.

   If your license has been granted, your name will be in the list, along with your call sign. Congratulations! You're licensed! Eventually, a paper copy of your license shows up in the mail, but you needn't wait for the hard copy in order to use the radio.

If you're using your GMRS radios for business, each employee that is not a member of your immediate family has to obtain his or her own GMRS license. Ouch. If you have a lot of employees, a business radio license (see Chapter 7) is more convenient, less expensive, and covers all of your employees.

Understanding Basic Radio Features

Every radio is a little different, but all are operated in similar ways and all the controls have similar names. The radios are a lot like DVD players — after you find out how to operate one DVD player, the rest are easy. This section uses the Uniden GMRS680 FRS/GMRS combo radio as an example. (Similar radios are available from Cobra (`www.cobra.com`), Motorola (`www.motorola.com`), and Audiovox (`www.audiovox.com`). If you like, you can download the actual operating manual for this radio at `uniden.com/pdf/GMRS680om.pdf`. You can follow along as each item is discussed.

Think generally. The radio you buy may have a knob where my example has a pushbutton or *key*. Figure 4-1 shows a typical radio's controls. Your radio may combine functions in different ways from the example shown. The important thing is to understand the control's function so that you can use it properly no matter what it's called or how it's shaped.

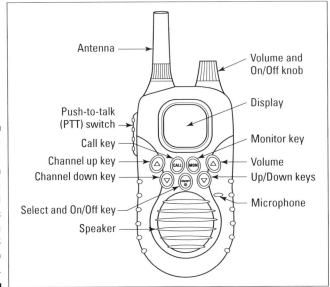

Figure 4-1:
The Uniden GMRS680 has the controls and features typical of a FRS/GMRS combo radio.

Images courtesy of Uniden Corporation

Operating controls

Here are some basic controls you ought to know about:

✔ **Volume and On/Off:** Some radios separate the volume and power controls, placing them on a *key* rather than on a combined knob. The power control may have its own button (usually labeled PWR or with that universal symbol of a circle with a line going through it) or it may be combined with another key. On the Uniden GMRS680 (refer to Figure 4-1), the power key is combined with the Select key. If your radio doesn't have a volume knob, it has two volume keys, one to increase and the other to decrease volume.

Turning down the volume does little to reduce the current drain on the batteries. To save your batteries, turn the radio completely off by using the power key.

✔ **Channel select keys:** Customarily, channel selection is made with two keys; one to increase and the other to decrease the channel number. The channels *wrap around* to start the sequence again; when you channel surf up the dial and reach 22, the next channel is 1; when you surf down the dial and reach 1, the next channel is 22, just like channel surfing on TV. Some radios use knobs to control channel selection.

✔ **Push to Talk (PTT):** To transmit, you must press a button on the side of the radio. Nearly all radios place this switch on the left side of the radio so that it can easily be pressed with the thumb (most users hold the radio in the left hand). The location is also convenient for using the index finger of the right hand. (Some radios have a voice-operated transmit mode, described in the section on VOX.)

Until you get used to PTT operation, you may forget to release the button when you're done speaking. Relax! That's normal. Just practice with a friend until you get the hang of it. Soon enough, PTT operation will seem completely natural.

✔ **Monitor:** Usually, your radio is quiet when no signal is present, no matter how high the Volume control is set. That is because of the *squelch* circuit (see the sidebar, "What is a squelch circuit?"), which prevents you from having to listen to a constant hiss and crackle of noise. A weak signal, however, may not be strong enough for the radio to turn on its speaker. The monitor key (usually labeled Mon) temporarily turns on the speaker no matter what the strength of the incoming signal. Use this when you want to check for someone calling or to listen to a weak signal. The MON key often is used to activate channel scanning (I discuss scanning in the section, "Indicators," later in this chapter).

✔ **Select:** Uniden uses the Sel label to indicate the key used to select options from menus, and most radios have a similar key to access menus. You can use the menus to navigate through (and adjust) various parameters and settings. Some settings include Privacy Codes, Call Tones, and more.

✔ **Call:** Pressing the Call key sends a selected ring tone. If the person or persons you want to talk to are within range and set to the right channel and privacy code, they will hear the tone. This is a good way to have several groups share a single channel. (See the section, "Rings and beeps," later in this chapter, for more information.)

Indicators

The largest indicator on the radio display shows the channel on which the radio is operating. This is the most important parameter, but there are several others you should know about. Check your radio's manual for the exact location and shape of each indicator.

✔ **Privacy Code:** This indicator shows which of the available codes you have selected, or indicates that you have not selected a code at all.

✔ **Transmit-Receive and Busy:** These indicators show whether you're transmitting or receiving and indicates whether a signal is being received. Some radios place these indicators on the display as a symbol or letter. Others use small colored lights. It's common for a red light to indicate that you're transmitting and a green light to illuminate when a signal is being received.

✔ **Scanning:** When scanning, an indicator appears to let you know that the radio is rapidly tuning through all the channels, pausing when it encounters a signal. Consult your manual for the exact function of your radio.

✔ **Call:** This indicator shows when your radio is transmitting a call tone.

✔ **VOX:** Intended for use with an accessory headset, the VOX indicator tells you when you're in VOX mode (VOX stands for *voice-operated transmit*). This indicator shows when VOX has been enabled. Basically, VOX replaces the PPT function mentioned in the previous section, enabling you to begin transmitting the second the radio picks up the sound of your voice.

VOX is very convenient when you are talking a lot, but if you're in a noisy or windy location it can be nothing less than a nightmare, causing your radio to transmit noise when you don't want it to. Find out how to adjust the sensitivity of the VOX feature and use the least sensitive setting to minimize unintended transmissions due to noise.

✔ **Battery level:** The battery indicator is a symbol is filled with fresh or fully charged batteries. It gradually empties as the batteries are drained. Watch this indicator, especially during transmissions. When the batteries are nearly dead, the battery indicator on many radios will blink.

✔ **Power level:** Some radios have a bar graph or other symbol that lengthens or gets darker with increased transmitter output power. Watch this indicator to help follow the rules about maximum power on FRS versus GMRS channels.

✔ **Lock:** With all those little keys, it's really easy to accidentally change a setting. This can be quite irritating! Thus, the Lock feature, which prevents any of the keys from changing things, was created. A symbol on the display shows when the radio is locked. Don't forget how to unlock the radio!

Introducing privacy codes

Along with the 22 channels of the FRS/GMRS services, nearly all radios also feature up to 38 different *privacy codes*. These are also known as *CTCSS tones, sub-audible tones, or PL tones.* (CTCSS stands for continuous tone-coded squelch system.) This system was invented to allow several groups to share a common radio channel, yet not have to listen to each other's messages.

What is a squelch circuit?

Even a radio guru gets tired of listening to the continuous hiss coming out of a radio with no signal present. The solution is a *squelch circuit,* sometimes just called *squelch.* Invented in the 1930s, a squelch circuit mutes the receiver output in the absence of a received signal, squelching (or silencing) the noise. The threshold at which the squelch *opens* (turns on the speaker) is usually adjustable. Raising the threshold so that a stronger signal is required to activate the speaker is called *tightening the* squelch. Lowering the threshold is called *loosening the squelch.* Exceeding the threshold to become audible is called *breaking squelch.* A marginal signal just barely exceeds the threshold; that signal is *just breaking squelch.* In order to hear these signals, which in many cases are noisy but quite intelligible, the Monitor button was invented to temporarily open the squelch. A squelch control is sometimes available to adjust the threshold.

The CTCSS uses low-frequency tones that are added to the voice of the user by the transmitter. The tones range from 67 Hz to 250.3 Hz and are generally too low in frequency to be heard by the listener or sustained in human speech. After a signal becomes strong enough to open the regular squelch system, a circuit listens for the selected tone. Unless the correct tone is present, the circuit prevents the signal from being heard, no matter how strong it is.

Privacy codes are very popular with FRS/GMRS users because they have only 22 channels to choose from. By using a privacy code, your family or group can leave the radio on all day and only hear transmissions from others on the same channel with the same privacy code. This reduces *radio chatter* dramatically and makes knowing when someone in your group is calling a lot easier. For example, if your family goes to the ballgame, you may notice that a lot of people are using the FRS channels. Trust me, you don't want to listen to them. If your group enables a privacy code (be sure you all get the same one!) on your radios, you can cut out all the other voices and just hear family members.

There are pitfalls, however:

✔ If someone in your group doesn't select the same privacy code as everyone else, numbered from 1 to 38, you won't be able to hear that person at all unless you turn privacy codes off. (A privacy code of 0 is generally used to turn off the CTCSS.)

✔ Your radio receives all signals, whether they have the matching privacy code or not. If a stronger signal without the matching privacy code is present, it overrides the weaker signal, but your radio remains silent. To avoid having to deal with mismatched privacy codes, only use them when you must.

If you are using a privacy code and your radio abruptly goes silent for a few seconds now and then, there is probably a strong signal present on the same channel with a different privacy code. To find out, turn off privacy codes so that you can hear every signal on the channel or use the Mon function. The easiest thing to do is to wait until the other users are finished or move to another channel.

Some of the fancier FRS/GMRS radios have the ability to listen to a signal and determine what privacy code, if any, is present. You can then enable the same privacy code and join the conversation. Radios use a variety of techniques to automatically look for and decode privacy tones.

A second type of privacy code is *digitally coded squelch* or *DCS*. When DCS is used, a short digital code is sent at the beginning of each transmission. If the receiver detects the proper code, it activates the speaker until the signal is no longer present. The combination of CTCSS and DCS gives a receiver a lot of control over what the user hears.

Rings and beeps

When you're sharing a busy channel, it can be aggravating to turn your attention to the radio every time a signal appears. The solution is *call tones.* If your radio supports this feature it might have up to five different tones that it can send. Call tones are similar to ring tones for mobile phones. By agreeing to use one of the five call tones, your group can be alerted to a call by the tone instead of having to listen to the beginning of every conversation. Call tones are selected in the same menu selection process that is used for changing channels and setting privacy codes.

The other type of tone your radio may generate is called a *courtesy beep*. When enabled, a short "peep" is transmitted when you release the PTT switch. This indicates to the listener that you are really done talking and not just at a loss for words. You must enable this feature through a menu selection.

Call tones and courtesy beeps sound (pun, pun) like a great idea, but use them sparingly. I once drove across Slovenia with a driver whose mobile phone was programmed to play *The Hungarian Polka.* By the end of the ride, we were both ready to never hear that, or any, polka ever again! Automatically generated noise gets old quickly and loses whatever urgency it was meant to have.

Basic Operating Skills

Learning about those mysterious buttons and knobs on the radio is just the beginning of your radio adventure. You still have to figure out how to drive.

It's not difficult at all, so read on, do a little practicing with a friend, and you'll be sounding like a seasoned pro in no time!

Holding the radio correctly

When I teach classes on radio use, it's easy to tell who the mobile phone users are. At first, they hold the radio against their heads between the ear and mouth. This won't work very well at all! Hold the radio so that the antenna is vertical, about 2 to 3 inches in front of your face, and turned slightly so that it is not facing you directly (see Figure 4-2). That keeps the microphone in the right spot and the speaker close enough to hear clearly. Keeping your antenna vertical is also very important — if everyone keeps the antenna upright your useable range is maximized.

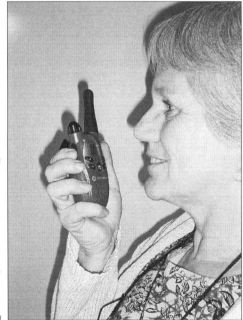

Figure 4-2: This photo shows the proper way to hold a radio. Keep the antenna vertical for maximum range.

Using a PTT radio

The biggest hurdle for all beginning radio users is unlearning telephone talking habits. Almost all radios can talk *or* listen, but *not* both at the same time. The users must take turns talking, unlike a telephone where both parties can (and often do) chatter simultaneously. You have to remember to press the

PTT switch to begin speaking (that's why it's called *Push to Talk*, of course) and then to release it when you're done.

You must also realize that the person talking can't hear you. He or she is uninterruptible! Pressing your own PTT switch while the person on the other end speaks has absolutely no effect. This reality is sometimes a problem with certain personalities.

Good radio practice dictates keeping your transmissions short for this very reason. It gives someone else a chance to break in. Consciously make yourself follow this rule and soon you'll do it automatically. Don't you wish you could make everybody do it?

Learning how to speak

Getting your message through on a radio requires clear speech. You can't mumble, talk with your mouth full, or rush your words. (Yes, Mom. . . .) The person at the other end can't see you! They can't read your body language or see you waving your hands for emphasis. Everything has to come through that little 1-inch speaker. Here are some simple rules that make communicating via any type of radio a lot more efficient:

- **Speak across (not directly into) the microphone:** Why? Place your hand in front of face and say, "Pepperpot." Feel those bursts of air with each "p?" This is why the radio should be turned slightly away from your face, so those airbursts don't overload the microphone and make a thumping noise in the receiver's speaker.

- **Use a normal tone of voice:** Guess what! YOU DON'T HAVE TO YELL INTO THE MICROPHONE! If fact, yelling makes audio quality a lot worse; when you're shouting to be heard, you're even harder to understand. Microphones can only respond clearly to a limited range of sound pressure. Speak too softly and background chatter or wind noise covers up your voice. Speak too loudly and your voice is distorted.

- **Don't forget to breathe.** If you find yourself getting tense and you hear the pitch of your voice rising, pause and take a deep breath. In fact, try to limit yourself to transmissions that are one breath long. This isn't the nightly newscast, so take the time to relax.

- **Your best friend is the phrase "Stand by."** Don't feel like you have to fill every second of radio time. The proper radio etiquette is to say, "Stand by" while you scratch your head. Don't waste battery power by saying, "Uhhhhhhhh . . ." until your get your thoughts together. Other listeners are expected to wait for you and won't jump in. When you are done speaking, say, "Over," or, "Go ahead," to cue the listener that it's his or her turn to speak.

✔ **Speak slowly.** Remember that all the visual clues we depend on in normal conversation don't exist on the radio. You may find that you have to speak at about two-thirds your usual speed (or even less if you're a *really* fast talker). If the person on the other end is writing down your message, slow *waaaay* down and imagine yourself writing each word as you say it. Speaking too fast often makes the overall process go slower than if you just let up on the accelerator a bit.

✔ **Take special pains to enunciate clearly, particularly words with lots of soft sounds, such as those made by the letters S, F, M, and N.** Letters and numbers that sound the same can also cause trouble. When pronouncing individual letters, such as when spelling a name, use phonetics (see the sidebar, "Hooked on phonetics"). Exaggerate the pronunciation of these numbers; say, "Thuh-ree, Foh-wer, Fi-yuhv, Ni-yun" if you don't want people to think you're saying, "Free, Floor, Fine, Mine."

Using a headset

Using a headset isn't much different than using a regular microphone, although you'll probably be able to hear better. Follow the same rule about not speaking directly into the microphone and keeping a normal voice. Remember that if you have headphones on, your instinct will be to speak too loudly when talking to others. If you use VOX (or voice-operated transmit) with the headset, breathing and throat clearing directly into the mic will activate or *trip* the VOX circuit just as speech will. If you can, reduce the sensitivity of the VOX circuit to the lowest level that still responds consistently to normal speech.

Scanning

Scanning is fun just for general interest. You'll be surprised and intrigued at the different conversations to be heard. It's also useful for trying to find active and inactive channels. If you are trying to find one particular person or group, however, it's not a very useful tool because the scanner stops for any and every signal. Most radios temporarily turn off privacy codes when scanning, so you'll hear everything. Check the radio's manual for information on the use of privacy codes while scanning.

Using call signs

Even if you are just using the FRS channels as an unlicensed user, using a call sign is a good idea under some circumstances. Casual users can just use their names, but if you are performing some kind of organized function, it's better to use the name of the function, instead. For example, when you call a business' main number and the receptionist answers the phone, you expect to hear, "Front Desk," or "Thanks for calling Acme Widget Company, this is Chris, can I help you?" You do not expect to hear, "Hi, this is Chris." Your group should develop a list of short call signs and then use them. For a parade, you might use "Judge's Stand," "First Aid," and "Staging." In a foot race, the call signs "Milepost 5," "Finish Line," and "Sag Wagon" might make sense.

Participating in public communications

Don't forget that everything you say is public. Everything. If you wouldn't say it to your mom, it doesn't belong on the radio. Some of the more capable radios have a voice scrambler option, if you need it.

Using Your Radio at Public Events and Places

The number one reason FRS/GMRS radios are so popular is to keep in touch with others in crowded or large areas. Lots of venues fit this description; shopping malls, sporting events, parades and festivals, and outdoor concerts are just a few locations where FRS/GMRS radios come in handy. Here are some guidelines for getting the best performance from your radios in those situations.

- ✔ **Choose the channel.** It sounds dumb, but choose the channel *before* you all go your separate ways.

- ✔ **Decide on a privacy code.** Do you really need a privacy code? When in doubt, do without, but go ahead and agree on a privacy code before you separate so that if you find you do need one everyone needs to turn it on at the same time, so be sure you're clear about when to turn it on (and off).

- ✔ **Test the radios first.** With the channel and privacy code options agreed on, do a quick radio and battery test before heading out. Make sure everyone knows how to operate the radio and change batteries.

- ✔ **Have a backup plan.** What if the original plan doesn't work or someone misunderstands? It's a good idea to have an alternate channel selected and a specific time to try it. Tell everyone, "Meet on Channel 18 without privacy codes at 2 p.m.," for example.

✔ **Be prepared for range problems.** Radios operating on the FRS channels with low output power won't be reliable for more than a ½ mile to 1 mile, much less indoors. Walls and floors can block radio signals. Be conservative in your expectations (see "Maximizing Your Range," later in this chapter).

If the channel you selected turns out to be too crowded, that's when you need the alternate channel. If you have privacy codes, your radio may act erratically in the presence of other signals. An alternate channel is probably the best choice. When operating at higher elevations, such as hilltops or on high buildings, nearby transmitters used by broadcasting and other radio services may overload the radio's receiver. You will probably have to move away from these locations to get reliable performance from the radio.

✔ **Keep fresh batteries with you.** Having a spare set of batteries is great insurance against a disaster. I prefer radios that accept AA or AAA cells (see Chapter 19) because I can get new ones anywhere and keep them in a plastic holder. Spare battery packs are usually available from the manufacturer. "Don't leave home without 'em!"

✔ **Protect your radio against theft and loss.** The good news is that radios are compact. The bad news is that makes them easy to lose or to swipe, especially when they're just clipped to a belt or strap. Many radios have a means to attach a tether — a good idea if you're going to be running around with your radio. To increase the chances of a lost or stolen radio being returned to you, engrave your driver's license and phone number inside the battery compartment.

✔ **Exercise courtesy.** Having a radio doesn't give you a license to be obnoxious. Talk in an appropriate tone of voice and only at appropriate times. Turn your radio off when a program is under way.

Using Your Radio in the Great Outdoors

After you get away from the crowds, FRS/GMRS radios are still very useful if they have adequate range. Don't expect your radio to "phone home" from the bottom of a ravine or the middle of the desert, but appropriately used, these radios can help keep a group together.

Skiing areas are a natural for a set of FRS/GMRS radios. You're almost always within range, even with low-power radios, except at the larger resorts. You may have to deal with interference from other users, but you can almost always find a channel to use. Ask the resort managers and guides what they suggest for radios.

Knowing the ARRL Wilderness protocol

In the backcountry or in the valleys that hold lakes and rivers, you can get seriously out of range of mobile phone service. Direct radio-to-radio communication is a sure way to find a lost or injured party. Ham radio operators have adopted the American Radio Relay League's (ARRL) Wilderness protocol to increase the chances of making contact with a lost or injured person while saving precious battery power. The protocol is to monitor certain frequencies for five minutes, beginning at specific times; 7 a.m., 10 a.m., 1 p.m., 4 p.m., and 7 p.m. A lost person is expected to transmit during those times and others are expected to listen, even if no one in their party is lost.

For example, listen from 7:00 to 7:05 a.m., using the monitor function to open the squelch so that weak signals can be heard. It is expected that radios would be turned off between these periods. This is a good reason to wear or have a watch with you! There is no standard FRS/GMRS channel nationally, but your group can use your regular communications channel and scan all channels for the same period, too. The simplest choice would be FRS channel 1 with privacy codes off. Hams use 146.52 MHz as a primary frequency and 52.525, 223.5, 446.0, and 1294.5 MHz as secondary frequencies. (Chapter 9 has more information on ham radio.) Even if you're not a ham, you can use the timing portion of the protocol.

Don't replace an avalanche safety radio with an FRS/GMRS radio!

Biking is another excellent application for radio communications, especially with a headset. Because you're out on the open road, you probably won't have any interference issues. The headsets keep both hands on the handlebars where they belong, but VOX can be a problem if it trips on panting and wind noise.

Hikers, hunters, campers, canoeists, fishers, and kayakers can all make good use of radio links because group members are frequently out of sight around a bend or down a trail. Manage your range expectations reasonably and don't let the radios give you such a false sense of security that you neglect trail and backcountry safety. You might consider adopting a version of the ARRL Wilderness protocol (see the accompanying sidebar), using a specific FRS/GMRS channel.

Power boating and off-road vehicles require higher-power GMRS capability because they can spread out over much longer distances. Nevertheless, using GMRS (licensed, of course) with an external antenna can provide reliable communications over several miles on a lake, river, or flat terrain. If you plan on traveling in rugged terrain or spreading out over more than a few miles, CB radio (see Chapter 5) is probably a better choice.

Maximizing Your Range

After you've become adept at configuring your radio to use the proper channels, codes, tones, and modes, you'll want to get the most range out of your radio. Simply play around with the radio and you'll find that indoor and outdoor ranges vary dramatically. Here are some ideas to help you get the most out of your radio and battery dollar.

✔ **Pay attention to your antenna orientation.** Make sure everyone is in the same habit of holding the radios so that the antennas point vertically. Random orientation of the antennas can lead to significant loss of range.

✔ **Move around.** Radio signals are blocked by or reflected off surfaces, hills, structures, large vehicles, and metal fences. When this happens, they interfere with each other, creating locations of low and high signal strength, just like water waves do. If you encounter poor signals, move the receiver and transmitter about ¼-wavelength (see Chapter 2), which for FRS/GMRS radios is about 6 inches. Just moving the radio around in the air while listening to the other station is sometimes enough to find a *hot spot* — a place with better radio reception.

✔ At the UHF frequencies used by FRS/GMRS radios, green foliage can absorb signals. Precipitation, including heavy fog, also takes its toll on radio signals. There's not much you can do except get in the clear from trees and brush. For weather-related effects, you'll just have to recalibrate your range expectations.

✔ **Look for higher ground.** If you can't get to a clear path to the other person, try getting to a higher spot, indoors or outdoors. Calling from a balcony or walkway indoors often helps. Outside, try walking up a hill or standing on a rock. You may even find yourself climbing on top of your car or up a tree.

Repeating Yourself

GMRS radios can use repeaters and FRS radios can't. Great. But what is a repeater? What does it do that is so valuable? Why are repeaters only available to GMRS radios? A *repeater* is a device that receives, amplifies, and retransmits a signal. Using a repeater with a relatively low-range radio such as a GMRS can significantly improve the range of your transmissions.

Repeaters communicate by listening on one frequency (the *input*) and at the same time transmitting what they hear on a different frequency (the *output*). Repeaters use what's called *duplex communications,* which means that they can receive and transmit on different frequencies at the same time. An FRS/GMRS radio uses *simplex communications*. This means that it transmits and receives on a single frequency, and it can't do both at the same time. Figure 4-3 shows how a repeater works. Regular radios can't act as repeaters because their circuitry can't transmit and listen at the same time.

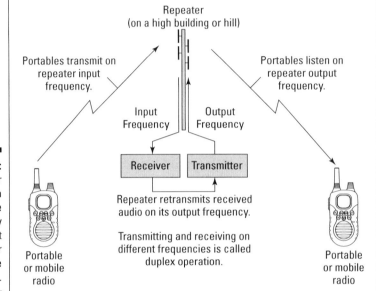

Figure 4-3: A repeater can listen on one frequency and transmit on another at the same time.

The difference between the repeater's input and output frequencies is called the *offset*. Refer to Table 4-2 to see the eight pairs of upper and lower frequencies. Each pair corresponds to one repeater channel or *pair*. The FCC has organized the GMRS channels so that there are eight possible repeater pairs to use, all with the same offset.

GMRS radios can also use simplex communications (direct radio to radio) on the *repeater output* frequencies (which are the lower frequencies of the two in a pair). For example, when using the "550" GMRS pair in Table 4-2, your radio would transmit on 467.550 MHz (the upper frequency of the pair) where the repeater is listening. This is called the *repeater input*. The repeater transmits on the lower frequency, or 462.550 MHz.

If you take a really close look at the table, you also notice that the repeater channels are all equally spaced from each other; the frequencies are always 25 kHz apart. This is called *channel spacing* and is illustrated in Figure 4-4.

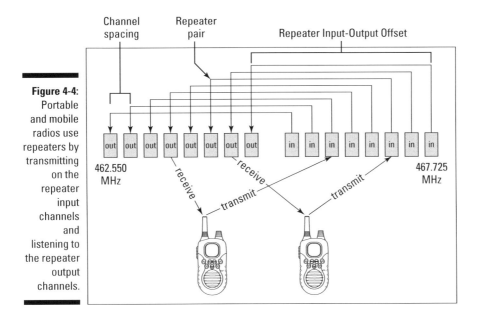

Figure 4-4:
Portable
and mobile
radios use
repeaters by
transmitting
on the
repeater
input
channels
and
listening to
the repeater
output
channels.

A repeater can't change the way your radio was built to function. So, a FRS/GMRS radio that was built to either receive or transmit (not both) at any given time will not suddenly be able to do both simultaneously. Which leaves open a very good question: If your radio can still only receive or transmit, why bother with a repeater? The answer is to increase range.

Repeaters are located in high spots on buildings, towers, or hills so that they can pick up signals over a wide area. They also have a powerful transmitter so that radios can hear them in the same area. This system allows two modest, handheld radios to communicate over a range of tens of miles!

As you might imagine, using repeaters is incredibly useful to businesses, such as delivery or messenger services. Most GMRS repeaters are owned and operated by business owners, although personal GMRS users are allowed to use them if they have the owner's permission. You can use the FCC's ULS Web site discussed earlier (see "Getting a GMRS License") to search for GMRS repeater owners in your area. Or get on your radio and ask via the repeater itself. If a control operator or the owner is listening, he or she may grant you permission.

Because repeaters provide service or *coverage* to wide areas, there are likely to be many users (or listeners) at any one time. Be considerate and keep your transmissions short.

Most radios do not have the ability to offset their transmit frequency and so can't use a GMRS repeater. This capability is found in radios intended for business and professional use, costing several times more than consumer-grade radios. If you have one of these radios, you can enable the use of repeaters by activating the *offset* or *repeater* function on the lower GMRS channels.

There is also another kind of repeater called *store-and-forward* or a *simplex repeater*. This type of repeater listens for a signal and then records its audio portion until the signal stops. When the signal stops, the repeater then re-transmits what it heard while the signal was present. This is less convenient than a real-time duplex repeater, but offers some functionality at lower cost.

Choosing a Radio

If you're ready to buy your FRS/GMRS radio, you may want to start by finding an extensive list of dealers for radios, antennas, and accessories. Try visiting the Open Directory Project's Radio Equipment category:

```
dmoz.org/Business/Telecommunications/Two-Way_Radio/Equipment/
                 Dealers/
```

Model numbers and features change very rapidly, so here's a simple guide to help you make your choice. The following are the three most important aspects of the radio.

Range: Specifications versus reality

Every radio package proudly announces the miles of range and bigger is surely better, right? Range for radios is like gas mileage ratings for cars. If all else is equal, the radio with the higher transmitter power probably has a higher range. An FRS user, limited to ½-watt of power, can only expect to get ½- to 1-mile range at best. Higher-powered GMRS radios can be reliable up to 2 or more miles, depending on terrain. Remember that the environment in which you use the radio affects useful range most of all.

Batteries: Buying packs or individual cells?

If you intend to use your radios for emergency communications, get the kind that that can use individual AA or AAA batteries — after all, these are the batteries you're most likely to have on hand or have easy access to. Easy access to relatively inexpensive batteries allows you to operate for an extended period of time if there is no ac power to run a charger. (You did remember to stock up on batteries, didn't you?)

If you're a casual FRS/GMRS user, the convenience of rechargeable battery packs or sealed radios outweighs the expense of individual cells.

With a little shopping around, you can get a radio that takes both recharge-able packs and the individual cells. If you use rechargeable individual cells (see Chapter 19 for information about batteries) you can have the best of both worlds! *NiMH* (nickel metal hydride) batteries are the best rechargeable cells and either NiMH or *Li-Ion* (lithium-ion) batteries make the best battery packs.

Warranty and reliability

The blister-packaged radios that come in pairs are not meant to be repaired. The store from which you purchased the package will replace the radios if they fail during the warranty period (and if you have hung on to your receipt), so shop for the longest warranty period you can find. The best manufacturers supply one-year warranties.

When it comes to reliability, radios can be hard to judge. Magazine reviews, such as those in *Popular Communications*, can give you some information, but rarely does the reviewer have the radio more than a few weeks. Clues that the radio is built to last are the longer warranty, heavier cases and rubber shock-absorbing panels, water resistance, and a wide operating temperature range. (If you plan on using the radio outside, be sure it doesn't stop working before you do!)

Another clue is whether the manufacturer has an extensive line of radio prod-ucts. Browse to the manufacturer's Web site. It will become obvious who the serious radio makers are. Their radios, in general, will perform better and more reliably. It's worth the extra few bucks!

Other options to consider

The following features have different value to different buyers. Pick the ones that are important to you and then start shopping!

✔ **VOX:** The convenience of VOX with a headset is a major plus when you want your hands to be free or you don't want to be pressing a PTT button all the time.

✔ **Memory channels:** Some radios allow you set up combinations of channels, privacy codes, call tones, squelch options, and so on, as *memories* or *memory channels*. This option can be very handy if you have several regular groups or functions that set up the radio the same way every time. Having a few memory channels keeps you from having to reconfigure the radio each time you use it for these regular functions. You won't need more than ten memory channels.

✔ **NOAA weather channels and SAME alerts:** You may already have a radio that can listen to the NOAA National Weather Service (NWS) weather alert broadcasts (visit www.nws.noaa.gov/nwr). Being able to use your FRS/GMRS radio to receive the local weather forecasts is a very useful combination. If the radio is also equipped to receive the SAME (Specific Area Message Encoding) alerts, even better!

✔ **Automatic ranging transponder system (ARTS or ATS):** This feature enables the radio to occasionally poll for other radios in a group to see if they are within range. If the other radio does not answer, an alert is sounded. This option can be great for keeping a family or group together.

✔ **Special packages:** Many radios now include a host of enhancements intended for use in specific activities. For example, several radios are available that include a Global Positioning System (GPS) receiver or a digital compass for use outside while hunting or hiking. Other models have clocks or can receive commercial AM and FM broadcasts. If you are going to use the radio in support of a specific hobby or sport, you might check with your favorite magazines or Web sites to see whether a manufacturer has put together a special package of features just for you.

GMRS-specific options

If you are going to get a GMRS license, consider whether you will want to use repeaters. Not all GMRS radios have the ability to use repeaters by offsetting their transmit frequency, so be sure your radio has this feature if you think you'll need it. Here are a couple of other things to look for in a GMRS radio:

✔ **Antennas:** GMRS radios can be used with external antennas. If you plan on using the radio while driving, removing the antenna mounted directly to the radio and using a mobile antenna instead can improve performance dramatically. Ask whether the antenna connector is a standard type. BNC- or SMA-style connectors are preferred because they are the most common. If the radio has a nonstandard connector, you'll need an adapter to connect an external antenna.

✔ **Battery packs:** The higher-power GMRS radios also consume batteries quickly, so be sure to evaluate the battery packs that are available. Purchase at least one spare. It's desirable to have a blank battery pack that you can load with individual cells for use when a charger isn't available.

Adding Antennas and Accessories

If you're using FRS radios, you won't be able to try out new antennas, but there are plenty of other ways to get more out of your radio and adapt it for various uses. This is a good opportunity to support your local radio shop, where you can try before you buy and see whether the gadgets that look good in the catalog actually work for your needs.

Antennas and cables

Changing the antenna on your radio from the short, flexible one provided with your radio to an external antenna improves performance more than any other single change. *Mobile* antennas, the most common external antennas made, are attached to a car. Antennas that are made to be installed on a building or mast are called *base* antennas. (Antennas, mounts, and cables are discussed in more detail in Chapter 18.)

To add an external mobile antenna, you'll need two items; the antenna and an antenna mount with an attached cable to connect to your radio. If you will be using the mobile antenna a lot, a semipermanent mount, such as trunk or *lip* mount with integral cable is best, because the connecting cable can be run inside the car and the antenna can be removed for security. Otherwise, use a clip-on or magnetic mount because you can easily remove it for any reason. These antennas usually require a one-piece assembly with antenna and cable.

The most common base antennas are *omnidirectional*, radiating equally well in all directions. *Beam* antennas, which transmit and receive best in one direction, are rarely required except in very difficult circumstances. You must purchase a beam antenna, cable, and mast or tower separately. (Remember that small GMRS stations are limited to 20 feet of antenna height.) When you purchase cable, be sure to get the right connectors installed on each end, or purchase the necessary adaptors. Dealers that sell antennas also sell cable, connectors, and other antenna accessories.

Microphones

You can also add a separate microphone to many radios. The most popular is a *speaker-mic* that contains both a speaker and a microphone with a cord 2 or 3 feet long that plugs into the radio. This setup allows you to mount the radio on your belt or in a charger while you use it. Pryme Radio Products (www.pryme.com) manufactures several versions of this useful accessory.

If you operate outside or in noisy or windy places, you may want to try a *throat microphone*. These are placed directly against your neck and pick up your voice through the skin. This eliminates quite a bit of noise, but they are a bit less convenient to use.

Headphones and headsets

If you want to listen without disturbing others or in a noisy place, earphones are used. While you can use full-sized earphones, a lightweight earpiece, or *ear bud,* is less cumbersome.

Combining earphones with a microphone creates a *headset.* These are light and comfortable. There are many variations suited to a wide variety of uses. Models are available with the PTT switch on the cable if your radio doesn't have the VOX feature. Radio dealers will carry a full line of headsets.

Bike and motorcycle accessories

Safe operating while pedaling or cruising down the open road requires that you have a method of using the radio that keeps your hands on the handle-bars as much as possible. To that end, *helmet headsets* are available that fit into an existing helmet or are integrated into a helmet. With a face guard to keep the wind out of your ears and away from the mike, communications to others in your group is a breeze, so to speak.

If you spend a lot of time biking or cycling, these accessories are worth investigating.

Battery chargers and power adapters

If you need to replace the charger that came with your radio or you want a spare, the dealer where you purchased the radio is the best source. If the radio has been discontinued, you can find surplus or spares on eBay.com. If you buy an after-market charger, check to be sure your radio and battery type are supported. Using the improper charger can ruin a battery. (Batteries and charging are covered in Chapter 19.) Power adapters for ac wall power or a vehicle's dc power must be compatible with your radio and battery. Check carefully to be sure that the power connector has the right polarity before hooking it up!

Chapter 5

Breaker, Breaker: Using Citizens Band

....

In This Chapter

▶ Introducing the merits of Citizens Band

▶ Using those knobs and switches

▶ Getting on the airwaves

▶ Choosing your radio equipment

▶ Operating your CB legally

....

During the great Citizens Band or *CB* craze of the late '70s and early '80s, it seemed like everybody just had to have a CB radio and the right jargon to sound just like a trucker. Movies like *Smokey and the Bandit* and television shows like *The Dukes of Hazzard* featured CB prominently; the airwaves crackled with colorful slang and ersatz truckers. While the fad days are over, CB radio is still alive and well.

CB is an unlicensed local radio service for individuals and small businesses. Radios are quite inexpensive and, used knowledgeably, can be very handy. In this chapter, I cover the various uses for CB radios. I also show you how to choose a CB radio. No chapter on CB radios would be complete without a discussion of antennas. If you have always wanted to get on the air and sound like a trucker, look no farther than this chapter!

CB Basics

How much of that stuff in the movies and on TV is for real? As you might expect, filmmakers and entertainers use a fair amount of artistic license in their portrayal of CB activities. If you start off your CB career with unrealistic expectations, you'll be disappointed, so I begin with the basics of Citizens Band. By the way, the name comes from the FCC's desire to encourage citizens to use the new radio technology; thus the Citizens Band was created in the early 1950s.

The CB service provides individuals and small businesses with a low-cost way to stay in touch and coordinate their day-to-day activities up to a 5- or 10-mile range. If this sounds like what you need, CB might be just right for you!

Getting help from experienced users

Learning the ins and outs of CB operating isn't always easy. Joining a CB club can help you over the sticky points everyone encounters at the beginning. Here are three places to go for more information:

- ✔ **REACT:** If you have any interest in emergency communication *(emcomm)*, REACT International (`www.reactintl.org`) has regional and municipal chapters of volunteers who are full of expertise; these people want to help newcomers. (REACT stands for Radio Emergency Associated Communications Teams, by the way.)

- ✔ **DXZone:** DXZone is an online resource for all amateur radio services, including CB. At DXZone, you can find a long list of CB clubs (`www.dxzone.com/catalog/CB_Radio/Clubs`).

- ✔ **Local Web search:** You may be surprised by the success you could have if you simply enter your city or region and the phrase *cb club* into an Internet search engine (for example, *"cb club" los angeles* yields over 300 useful results).

  Surrounding the phrase with quotation marks — *"cb club"* — filters out results that don't have anything to with CB radio.

- ✔ **Reference sites:** You can get a lot of great background information from reference Web sites, such as `www.popularwireless.com/cbweb/cbwords.html`.

  Most CB clubs are online groups that offer advice and answers to questions. (Be wary of clubs advocating the illegal styles of operating listed at the end of the chapter.)

Getting the lowdown on licenses and requirements

Many two-way radio services require you to purchase a license from the FCC. The CB radio service isn't one of them. The FCC does have some rules that you *should* follow, though. You can find the rules tucked into the operating manual of your radio (you did keep the operating manual, didn't you?). You can also download them from `www.reactintl.org/rules-cb.htm`. The rules are easy to read and are organized as a list of common questions. They include technical rules about radios and antennas, as well as what you can and can't do on the air.

What you *can't* do is fairly straightforward and laid out clearly in CB Rule 13. You can't advertise materials for sale or a political campaign, cuss, play music, or rebroadcast radio or TV programming. You're also forbidden to intentionally interfere with other stations (duh!) and make false transmissions, particularly distress calls. FCC rules also prohibit contacting stations in other countries (except Canada).

Knowing CB frequencies and channels

Each CB radio can operate on any of 40 channels centered on frequencies from 26.965 to 27.405 MHz. This frequency range is near the upper limit of the traditional shortwave or *HF* (high frequency) band, tucked in between the Amateur Radio 10-meter band and a band used by business radios.

You can operate in regular AM mode (see Chapter 2), or select either upper or lower *sideband,* a variation of AM discussed later in this chapter. That gives you 120 different choices about where to operate!

Finding Uses for Your CB Radio

By far, the most popular use of CB today is in vehicles. Using CB for business-to-business communications is less common than it was in the halcyon days of CB, but farms, towing companies, local delivery services, taxis, and other mobile users still find CB very useful.

Professional drivers (truckers, taxi drivers, and so on) use CB radio for everything from keeping an eye on traffic (and speed traps) to checking in at the delivery dock and making idle conversation with anyone in range. Right behind the professionals are private citizens, just like you, who use CB when they drive for many of the same reasons.

Because so many individuals have CB radios, it's common that the businesses they frequent will use CB to talk to them. For example, an RV park or truck stop might *monitor* one of the channels all the time to answer questions and instruct drivers. If you look closely at billboards and other signs advertising road or recreational services, you might see a channel number showing where to tune.

The good news is that Citizens Band, being a public, unlicensed service, is open to everyone. The bad news is that it's open to everyone. Sooner or later, you'll encounter someone being rude or offensive. Sigh. There's not much you can do except tune to another channel or ignore the offender. Getting into an argument serves no purpose. Have a backup channel and don't hesitate to use it!

Getting To Know Your Radio

The following sections give you the basics about what the controls and indicators on your CB radio do. I use one of the simpler radios, the Uniden PRO520XL, shown in Figure 5-1, as an example. This model is typically used in the car on a road trip. You can find fancier models, but all models have these same basic functions.

In the following sections I assume that you've already installed your radio, attached an antenna, and that everything is working. If you need help with installation, see the technical chapters in Part IV and check your operating manual.

Under control: Knobs and switches

No matter what kind of radio you have, you'll have keys, knobs, switches, and buttons galore. All the following items are located somewhere on your radio, and shown in Figure 5-1:

- ✔ **Volume:** Sometimes called AF *(Audio Frequency)* Gain, this controls the volume of the received audio.

- ✔ **Squelch:** This sets the level of received signal required before audio is heard from the speaker. (See Chapter 4 for a complete description of the squelch control.) Turn the control to the right to raise the level.

- ✔ **RF gain:** Radio Frequency gain controls the sensitivity of the receiver to incoming signals. Reduce sensitivity in the presence of strong signals.

- ✔ **Microphone connector:** Be sure to use a microphone compatible with your radio. The 4-pin Amphenol connector in the figure is the most common style, but you may find 5- and 8-pin DIN, and RJ-45 modular connectors.

- ✔ **PTT:** Press the Push to Talk switch (located on the microphone) to activate or *key* the transmitter to put your signal on the air.

- ✔ **Channel selector:** Use this knob to tune the radio to the operating channel.

- ✔ **PA:** PA means *public address.* When you set your radio to PA, instead of transmitting a radio signal, your voice is sent to a speaker plugged into the PA speaker jack on the back of the radio.

- ✔ **ANL or NB:** Automatic Noise Limiting or Noise Blanker activates a circuit in the radio that clamps the audio from static crashes and other noise pulses, which might otherwise cover up the received station's signal. Strong signals can confuse the circuit and distort the received signal, so turn ANL off when you don't need it.

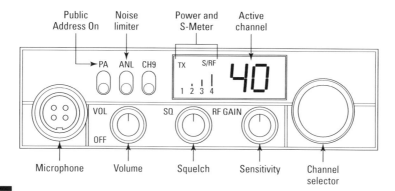

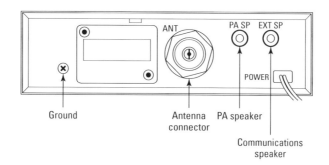

Figure 5-1:
Front panel
controls and
back panel
connections
of a CB
radio.

Keeping an eye on indicators

You could have the most basic radio or the most complex. In either case, the following indicators are somewhere on the radio:

- ✔ **Power:** Either a meter or series of lights shows the level of transmitter output power. The lights or meter increases in power when you increase your voice level. The meter may also have an SWR (standing wave ratio) scale, which I discuss later in this chapter.

- ✔ **S-Meter:** The *S* refers to *S-Units,* which measure signal strength. Unless your radio's meter is calibrated in S-units, received signal strengths are just relative. A signal that pushes the S-meter to 9 units, or S-9, is a strong signal.

- ✔ **Active Channel:** This displays the channel on which the radio is operating.

Using CB outside the United States

A number of countries have Citizens Band licenses. In Canada, CB is referred to as the General Radio Service, but it works pretty much the same way. Some countries don't use the same channel frequencies as the United States. For example, in the United Kingdom, CB channels span from 27.6 to 28 MHz — above the U.S.'s Channel 40. Because of the way radio signals propagate, it's unusual for CB signals to span the oceans. CB is designed to be a short-range service, anyway.

Along with the HF channels, some countries also have what is referred to as *UHF CB* or ultra high frequency CB. The United Kingdom allows CB operation on channels around 934 MHz and Australia has 40 channels of CB operation at 477 MHz. These services use FM signals and are very similar to the American FRS/GMRS service discussed in Chapter 4.

Depending on the model and brand of radio you're using, your radio may also have the following indicators:

- **RX/TX:** This indicator lets you know whether your radio is receiving or transmitting. (RX means you're receiving; TX means you're transmitting.)

- **Antenna Warning or SWR (standing wave ratio):** This indicator lets you know that something is wrong with your antenna or connecting cable. Stop transmitting and fix the problem. If your radio has an SWR meter, you can read SWR as a ratio of 1:1 or higher. SWR is an indication of how well your feedline and antenna are accepting the transmitter power output. Values of 2:1 or higher indicate a problem.

Getting a gander at the back panel

On the back panel of every radio, you find the following:

- **ANT:** The connection for the antenna. The screw-on connector on the radio is an SO-239. The mating cable connector is a PL-259.

- **EXT or COMM speaker:** Your radio probably has a small internal speaker, but for better audio quality, you can connect an external speaker here. This cuts off the internal speaker.

- **PA speaker:** You can connect a speaker for your voice to be heard by people outside the car. Typically, a small PA *horn* speaker is mounted under the hood of the car. If you have the PA speaker too close to the microphone, you'll have feedback.

✔ **Ground:** Connect the radio's metal enclosure or *chassis* to the car's frame (or a ground rod at home) with a thick wire or copper strap to prevent your signal from feeding back into your microphone.

Little extras for higher-end radios

The following controls not shown in Figure 5-1 are found on radios with more features:

✔ **Sideband:** You can operate with AM signals or upper *(USB)* or lower *(LSB)* sideband, which I describe later in the chapter. In either of the sideband positions, you can receive AM signals. In AM position or if you select the wrong sideband, received sideband signals sound like Donald Duck.

✔ **Mic or Mic Gain:** Turning this control clockwise increases the amount of voice signal from your microphone (or *mic*). Turned up too far, your voice is distorted. You may cause interference on adjacent channels, as well. Turned down too far, your signal is weak. The proper setting is so that the power meter follows your voice level, showing maximum output power just during the peaks.

✔ **Clarifier or Delta Tune:** Not all radios transmit exactly on the correct frequency. The clarifier control acts like a fine-tune control to shift the receiver to receive signals properly. Clarifier does not affect your transmitted signal.

AM versus single sideband

In Chapter 2, I detail the difference between AM and FM signals. Sideband signals are a variation of AM signals. An AM signal is composed of three components: a carrier in the middle, plus bands of RF energy, above and below the carrier frequency, called *sidebands*. The carrier is just a steady signal, where amplitude and frequency never vary. The sidebands appear whenever you speak into the microphone; they're as far away from the carrier frequency as the frequencies in your voice. For example, if you whistle a nice clear toot at 1,000 Hz, sidebands appear 1,000 Hz above and below the carrier.

Why are sideband signals used? Because the carrier doesn't change, it doesn't carry any information. Furthermore, both sidebands carry identical information. A *single sideband,* or *SSB,* signal does away with the carrier and one of the sidebands, leaving either the upper sideband (USB) signal or lower sideband (LSB) signal. (A USB signal is shown in the figure.) This allows all the output power to go into the signal that carries your voice, increasing your range.

✔ **Distant/Local:** In the Local position, the received signal level is reduced (*attenuated*) to prevent a strong signal from overloading the receiver.

✔ **WX or Weather:** This control tunes your receiver so that you can listen to one of the VHF weather stations operated by the National Weather Service.

✔ **Channel 9 or Channel 19:** This control automatically switches your radio to the emergency channel (Channel 9) or traffic channel (Channel 19).

Your microphone may also have push-button controls that enable you to change channels up and down or adjust microphone audio output level.

Operating Your CB

Here you are, smelling that new-radio smell, ready to get on the air! You may be wondering what you should expect. You may be unsure of what to say. Well, never fear; the following sections give you the basics about listening and transmitting. I start with listening — always a good first step in an unfamiliar environment.

Receiving your first CB transmissions

First, open the squelch (turn the squelch control all the way counter-clockwise or to the minimum setting) and set the volume to a comfortable level. If you have used an FRS/GMRS or business band radio before, you may be used to crystal-clear transmissions, or, at worst, the hiss of an FM radio signal. Not so with CB radios — don't be surprised to hear popping and crackling noises rather than the hiss of an FM radio.

Tune to different channels and see who you can hear or *copy*. To get a good idea of range, listen to Channel 19, the highway traffic channel, and try to figure out where the stations are. If you hear someone say, "mile marker," the number tells you where the person is on an interstate highway. Listen to the stations until they fade out (when people travel out of range, you've *lost them*). You can hear stations giving their *10-20* or *20,* which means location.

Whether a signal is loud or weak also depends on the height and quality of the antenna installation. Soon, you'll have an educated opinion on the range of your CB.

Handling noises and interference

Inevitably, you will hear various kinds of interference when you use your CB radio. When a weak station is covered up by a stronger one, it's called *co-channel* or *on-channel* interference. You may hear grumbly or scratchy noises that turn out to be synchronized with a signal on a nearby channel. This type of interference can be caused by a poor-quality signal or (if the offending signal is very strong) by overload in your receiver. The interference is called *splatter* for obvious reasons.

If you're in the car, you might hear a crackle that changes pitch with engine speed. At home, you might hear noise from a car's ignition system as it goes by or the CB might pick up a motor in your home, such as a sewing machine or furnace blower. There may be short crashes caused by storms. These are the sounds of the radio environment around you. In contrast, FM radio suppresses these noises, but requires more power and frequencies to transmit.

Sometimes you'll hear a steady whistling sound that's caused by two signal carriers combining in your receiver, generating an audio tone equal in frequency to the difference in frequency between them. This is called a *heterodyne*. You can also hear heterodynes on the AM broadcast band at night when a distant station's carrier combines with one from a nearby station.

A skip and a hop

Through the quirks of radio propagation described in Chapters 2 and 15, you will occasionally hear signals from a long way off. It's not terribly unusual for CB signals to make the journey across a continent or even between continents, depending on how the sun is feeling. What's going on here? CB radio uses frequencies that are at the upper edge of the ionosphere's ability to turn them back to earth. A radio signal's trip up and back to earth again is called a *hop;* the process of reflection is called *skip.* Signals can even make more than one hop, enabling communications over thousands of miles.

On a day with lots of solar ultraviolet shining your way, you may hear dozens of signals from hundreds of miles away. As this book is written (2005), the sun's *sunspot cycle* (www.copper.com/sunspot.html) is at low ebb; this means that the CB bands are populated just by local signals. Making contact over these long distances is known as *working skip.* Although it sounds like a lot of fun to communicate across the miles, the CB rules prohibit it. If you hear stations making contact this way and you can't resist taking part in the fun, why not check out getting an amateur radio license (see Chapter 9)? Working skip is not only okay with hams, it's encouraged!

For usual operating, *tighten the squelch* by turning the knob clockwise or increasing the setting until the noise is cut off. Noises may break through from time to time, but the idea is to get rid of most of it. You'll get good at setting the squelch level as you get used to the noise in your area.

Making your initial communications test

You can listen all you want, but why not do some talking? When you find a clear channel to do your initial test, listen for at least 10 or 15 seconds to be sure someone isn't using the channel. When you think you have a clear channel, follow these steps:

1. **Press the PTT switch and say, "Is the channel clear?" into the microphone. If someone says the channel is busy, just tune to another channel — no need for further transmissions.**

   You should speak *across* the microphone at an angle and use a normal tone of voice. (Chapter 4 contains a figure of proper speech technique.) When you speak across the mic, the puffs of air in your voice don't cause as many pops and thumps in your transmitted signal.

2. **Give a *five count* by saying "Testing, one, two, three, four, five."**

   While you speak, watch the transmitted output power meter and adjust your Mic Gain or speech level so that the output just hits peak power at the loudest points of your speech.

   If you turn up the mic gain much more, your voice may become distorted. Watch the transmitted power meter while you talk until you get more comfortable on the air.

Picking a handle

Before you spend any time chatting up your CB buddies, you need to pick a *handle* — your radio nickname. If you've spent any time on the Internet, you'll already have a User ID for e-mail and Web sites. You may have one or more chat room names, too. Your CB handle is exactly the same except that you say it instead of typing it.

Even a short period of listening will give you plenty of ideas about handles. Here's your chance to be *Rough Rascal* or *Pirate Gal.* You can change any time, so if you get tired of one handle, try another. You can remain relatively anonymous with your handle, which may help you overcome mic fright. Don't pick a suggestive or naughty handle — unless you want that kind of attention — and don't pick a complicated one. Keep your handle short and simple.

The term handle is an old slang term for name that goes back to the cowboys of the Old West. Telegraphers picked it up and the ham radio operators got it from them. CB operators copied the hams, and there you have it. Have you picked one out yet?

Learning communications basics

When you feel relatively comfortable with listening in, you have made your first channel test, and you have a handle (I'm calling you Clean Jean), you're ready to take the plunge. You get to the savvy-sounding jargon later, but here are some samples to get you rolling:

- ✔ **Radio Check:** Asking for someone to comment on your signal quality and strength is called getting a radio check. Either call on a clear channel or break in to an existing conversation and say, "This is Clean Jean for a radio check, please."

- ✔ **Making a Call:** If you hear someone you want to talk to (say he calls himself Rubber Ducky), call him when the channel isn't busy and say, "Rubber Ducky, this is Clean Jean. Do you copy?" If Rubber Ducky hears you, he'll respond.

- ✔ **Breaking In:** If you want to enter an ongoing conversation, wait for a pause and quickly say, "Break." Don't be wordy, just let the other operators know there's someone else out there. You may have to try a couple of times before they hear you.

  Don't say, "Break, break," twice in succession, because that means "Emergency" in many areas.

- ✔ **Traffic Check:** Say you want to know how traffic is on your morning commute and you're headed into town eastbound on I-80 at exit 42. You'd say, "This is Clean Jean eastbound on I-80 at exit 42. How's traffic downtown this morning? Over."

- ✔ **Signing Off:** To conclude a conversation and let others know the channel is clear, say, "This is Clean Jean, clear," or, "Clean Jean is clear."

  Don't say, "Over," at the end of a conversation. "Over" is what you say between sentences during a conversation.

There are lots and lots more ways to say just about anything on CB. The service has built up quite a lot of colorful jargon and slang — far too much to list here. However, if you want to master the lingo, you can find several good lists of definitions on the Web, such as www.popularwireless.com/cbweb/cbwords.html. CB lingo is a living language and varies strongly by region, so you won't hear the same slang in Nashville that you do in Los Angeles, London, or Melbourne, mate!

Don't ladle too much jargon into your transmissions. You'll just be viewed as a wannabe and not taken very seriously. Use the jargon as you get comfortable and don't try to put on an act.

Going Out and About with Your CB

CB radio is a great communication tool when you're out driving the roads or on the water. In many places, you may find yourself out of range on the VHF/UHF channels used by FRS/GMRS radios. CB extends that range by a few miles, covering a lot more road, lake, or trail. The small size and low height of mobile antennas often result in compromised performance, however, so be sure to read Part V to get the most out of your radio!

By far, the most likely use of a CB is in a vehicle for sharing traffic information and for emergency communications on the road. In fact, the FCC has designated Channel 9 as the emergency and traveler assistance channel. You are not supposed to chat on Channel 9. You shouldn't use it to try to quickly call a friend so that you can move to another channel and chat. It's worth noting that many police cars have a CB radio in the patrol cars, monitoring Channel 9.

Many CB radios have a Channel 9 switch that overrides the main channel selection control, setting the radio directly to Channel 9 in a hurry. Some radios also feature a "Channel 9 Priority Watch" that lets you listen to another channel, but jumps to Channel 9 whenever a signal appears there.

If getting traffic information, hearing about jams, accidents, road conditions, and the location of officers of the law is your cup of tea, you'll find this info on Channel 19, also known as the Trucker's Channel. Channel 19 is where you'll find the CB radio made popular on the silver screen and in the songs of C.W. McCall.

Because having everyone in a large group use a single channel would lead to horrible congestion, *calling channels* are often used. To make a contact, you call the other station on the calling channel and, if the person you're looking for responds, you both agree to move to a less-populated channel. That's a good way to find each other quickly and reliably. Don't hog the calling channel by chatting there — you'll aggravate everyone within earshot.

Motorcyclists congregate on Channel 1, moving up to Channels 2, 3, or 4, depending on radio congestion. (They don't want to take their eyes off the road to be dialing around.) You'll hear chatter between groups taking a day cruise or lone cyclists on a long trip. Channel 4 is also popular with off-road vehicle enthusiasts and is used to coordinate races and driving events.

As you pass an RV pulling up a long grade, take a good look on the back of the vehicle for a small sign listing of a CB channel. There's a good chance the driver is listening there. RV clubs encourage using Channel 13 (Channel 14 in

Europe for *caravanners*) as the calling channel. At a campground, look to see if the manager's office uses a CB channel for local communications, too.

While fishing or boating, you'll find Channel 3 to be a popular watering hole for mariners. Marinas and harbor services often monitor Channel 3 as a boater's intercom. (They also monitor marine VHF Channel 16, as I describe in Chapter 8.)

Using Your CB for Emergency Communications

Many people buy a radio and toss it in the car or boat thinking that it will be their personal 911 service in case of car trouble. Without a little preparation and planning, the results are likely to be disappointing. Without practice, how would you know how to communicate in an emergency? You also need to know whether your equipment is working, and the best way to do so is to use it periodically. Use your radio from time to time and listen regularly so that you get a good idea of what to expect on the air.

If you need help, follow these steps:

1. **Tune to Channel 9.**

2. **If the channel is clear, say, "This is [your handle], and I need assistance."**

3. **Wait 10 to 15 seconds and call again if no one responds.**

   Check your meter to be sure your transmitter is working.

4. **When you get a response, report your location, the nature of the problem, and what kind of assistance you need.**

5. **If no one responds, tune to another channel, such as 19, listening for a station with a good signal.**

6. **During a pause in the conversation, say, "Break, break!" to indicate that you have an emergency. If someone hears you, you'll get a response.**

Every transmission you make is public and can be received by anyone within range. When you make a call for help using any public radio channel, remember to:

✔ Stay in the car and lock the doors — you're much safer there than you are walking around.

✔ If someone arrives at your car, roll down the window a little bit and ask the person to contact the police or a roadside service company for you.

✔ If the person claims to be a public safety officer, ask to see identification.

✔ Never get into a stranger's car.

Shopping on the CB Channel

To get on the air from your car, you need three things; a radio, an antenna, and a cable to connect them. To operate from home, you need all three of these items, plus a power supply for the radio. You can purchase these items as a package or get them separately. (All new CB radios you might purchase come with a microphone.) Figure 5-2 shows you the basic pieces you'll need to set up your CB station.

Radios, or *rigs,* come in two basic flavors; *mini rigs* and *base quality.* Both have the maximum allowable output power:

✔ **Mini rigs** include handheld and compact models. They have the minimum number of controls and are intended to be easily transported and installed temporarily in a vehicle or used while hiking or camping. These rigs will get you on the air but generally omit noise limiting and other nice-to-have features like weather or priority channels.

Good examples of compact radios include the Uniden PRO510 and RadioShack TRC-503. The Cobra HH38WXST is a high-quality hand-held CB — not a toy walkie-talkie.

✔ **Base-quality** radios are suitable for mobile or home installation and have a complete set of controls, better and larger meters, and several nice-to-have features. Sideband operation (which often comes with base-quality radios) is nice to have, but it isn't necessary. In fact, I recommend that you operate using AM for a while to see if you really need sideband.

If you plan on operating from home or permanently installing a CB in your car or RV, get a base-quality rig. The extra features and solid construction will serve you well. These rigs usually include a better-quality microphone, too. Examples of base-quality radios include the Uniden PC78ELITE, Galaxy DX 949, and Cobra 75WXST.

If your radio requires 12 V dc power, you'll have to buy a power supply to operate it from 115 V ac power. Check your radio's operating manual for the maximum amount of current draw. The power supply should have a rating of at least 125 percent of the radio's current needs.

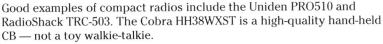

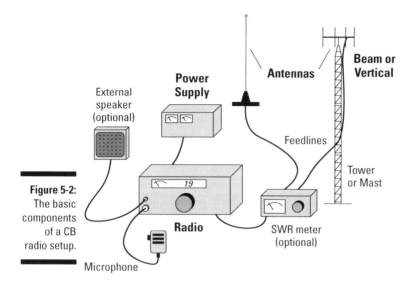

Figure 5-2:
The basic
components
of a CB
radio setup.

Understanding your SWR meter

Figure 5-2 shows a couple of optional accessories, the external speaker and the *SWR meter*, sometimes called a *SWR bridge*. SWR stands for *standing wave ratio*, which is a measure of how well your antenna is tuned. A properly tuned antenna accepts all the power from your radio and radiates it out into space as radio signals. If the antenna isn't tuned exactly right, some of the power bounces back and forth in the feedline between the radio and antenna until it is radiated or lost as heat in the feedline. When the antenna is adjusted just right, the SWR ratio is 1-to-1, or 1:1.

Typically, an antenna can only be perfectly adjusted at one frequency or on one channel. When tuned up on Channel 19, most work fine on all 40 CB channels without readjusting or *retuning*. You should be concerned if the antenna SWR exceeds 2:1. SWR higher than 2:1 means your antenna needs an adjustment. It may also indicate a feedline or antenna mounting problem.

Sounding great: Microphone madness

Because your radio's power output is relatively limited, CB users are constantly looking for ways to make their signal more powerful and understandable. The natural tendency is to turn your mic gain knob all the way up or start yelling when the other station doesn't quite copy your transmissions. All you accomplish is to add a lot of distortion to your signal and create interference on adjacent channels. Don't do it!

A *power mic* is a microphone that has a battery-powered amplifier to boost the signal applied to your rig. A little boost goes a long way, and it's easy to tell when an operator has overdone it. Modern rigs are plenty sensitive and really don't need the higher audio levels to get to full power. *Speech compressors* and *processors* are used to increase the average power of a signal by amplifying quiet sounds more than loud ones. A modest amount of compression can help you cut through noise and interference. Note that I said *modest.* Too much compression causes your signal to become noise and interference! Compressors should never be used outdoors or in a moving vehicle because they will happily amplify all background noise right along with your speech, making hearing and understanding you harder for others, not easier.

The very best way to adjust your audio accessories is to be on the other end and listen to yourself. Because that's rather difficult, you need a radio buddy to listen to your signal and tell you, "That's too much!" When you get your signal sounding good, take note of the control settings and how your power meter reacts to your voice. Then you'll know how everything is supposed to look when your signal is at its best quality.

The way to better audio quality and crisper, punchier signals is to use a high-quality microphone. Cobra (www.cobra.com) has models that perform noise canceling (great for mobile use!) with excellent audio response. Astatic D-104 Silver Eagle desk microphones are available from numerous dealers and have a well-deserved reputation for excellence. Spending a little extra money on a quality microphone is money well spent. When you hear a station on the air with an excellent quality signal, be sure to ask what equipment the caller is using. Maybe you'll get such a request yourself some day!

When you hear a beep or bloop at the end of a sentence, that's known as a *roger beep.* First used by NASA, the beep at the end of speech signals that the station has released the microphone PTT switch and takes the place of saying, "Over." This helps prevent *doubling* — two stations speaking at the same time. Some radios include this capability; or you can buy a microphone that provides its own beep. The roger beep is handy when several stations are having a conversation. In casual operation, turn it off when you don't need it.

Many stations have also added *tricks boxes* that add all sorts of sound effects and echo to the speech. There's no rule against them, but they don't do a lot for intelligibility. It's not for nothing that stations using such gadgetry are often referred to as *space cadets.*

Choosing and using antennas

There are two basic types of antennas used with Citizens Band radios: *verticals* and *beams.* Verticals receive and transmit equally well in all horizontal directions. They are used both at base stations and by mobile operators.

Beams, on the other hand, are so-named because they direct or *beam* their energy in a preferred direction. Beams are much larger than verticals and require a mast and rotator to change directions (see Chapter 18). For a much more detailed discussion of antennas and SWR, browse to `electronics.howstuffworks.com/question490.htm` or `www.signalengineering.com/ultimate`. If you are interested in antennas, the FCC permits you to design and build your own. I really enjoy doing so!

Using a vertical antenna

You can't get much simpler than a vertical antenna, electrically or mechanically. As a result, these antennas tend to be sturdy and reliable, standing up to the elements, even in exposed locations. Verticals are also easy to mount to masts. Sometimes referred to as *ground-plane* antennas, the simplest form of a vertical antenna is a ¼-wavelength of wire or tubing with several *radial* wires attached at the base.

The simplest verticals have just one wire or tube (called an *element*) that does the radiating. More complex designs combine several radiating elements stacked above or beside each other to increase gain. These are common types of vertical antennas:

- **Base:** A vertical antenna designed to be installed at a fixed location.

- **Whip:** A vertical antenna with a single, continuous radiating element.

- **Collinear:** An antenna with several elements stacked vertically, focusing more energy broadside to the antenna.

- **Loaded:** A vertical antenna with small coils inserted in the radiating elements to decrease their physical length.

- **Rubber duck:** A flexible, plastic-coated antenna used with handheld radios.

- **⅝-wave:** A vertical antenna with a single radiating element ⅝-wavelength long.

Most vertical antennas for home use are designed for mounting on a pipe or tube and to be connected to the radio with coaxial cable. Unless your operating manual specifically says otherwise, it does not matter whether the support is metal or an insulating material such as fiberglass, plastic, or wood. Avoid mounting antennas above a chimney, if possible, because the soot and combustion gases will lead to an early end of life for the antenna and corrode the connection to the feedline.

Using beam antennas

A beam antenna's elements can be aligned horizontally or vertically. If you want to communicate with stations that use vertical antennas, as is the case with most CB stations, your antenna's elements should be vertical, too.

Is the gain plain?

Gain is the measure of an antenna's ability to transmit or receive signals in a certain direction. An antenna with gain in a preferred direction is said to be *directive* or to have *directivity*. A drawing showing the antenna's gain in any direction is called an antenna's *radiation pattern*. Similarly, the ability of the antenna to reject signals in unwanted directions is called the *front-to-back* or *front-to-side* ratio.

Gain is measured in *decibels*, abbreviated as dB. Because a vertical antenna radiates and receives equally well in all directions, it has zero dB gain, and its radiation pattern is a circle. Beams may have gains of anywhere from a 3 dB (essentially doubling your *effective* power) to 8 or 9 dB, which is the same as multiplying your power by 8! Don't be fooled by extravagant gain claims. The best way to assess it realistically is to ask what the gain of the antenna is compared to a vertical antenna.

Beam antennas with straight elements are called *Yagis* (after their inventor). A variation of the Yagi substitutes loops of wire for the straight elements. Because these loops are usually four sided for mechanical convenience, these antennas are called *quads*. See Chapter 18 for more information about mounting beam antennas.

Operating on the Right Side of the Law

Citizens Band being the public, unlicensed service that it is, there is a certain element of its users that thrive on boorish behavior and breaking the rules. In the hope that a list of common transgressions won't be read as an instruction to do them, here's some definite CB don'ts:

- ✔ **Don't use illegal amplifiers:** Also known as *kickers* or *foot warmers,* they are used by illegal power stations and often overwhelm anyone else on the channel. They cause interference for several channels on either side.

- ✔ **Don't work skip:** It's technically illegal, although it is not as disruptive as using an overamplified station.

- ✔ **Don't make illegal radio modifications:** Modifying or *opening up* a radio voids the type-acceptance and is likely to upset its functions, too. Beware of the so-called combo radios (called CB/Ham radios); they're not legal in the U.S.

✔ **Don't freeband:** Operating outside the CB bands, usually with an illegally modified radio, is called *freebanding*. Although there's quite a number of people doing it and even assigning themselves made-up call signs, this is a highly illegal activity.

Does the FCC really care about illegal CB operation? For minor transgressions and day-to-day poor operating, not really. The people at the FCC are *generally* too busy to care. Note that I said *generally;* there are no guarantees in radio or in life. Folks that thought they could slide up into the nearby 10-meter amateur band and manufacturers of easily modified (wink, wink) radios that can operate outside the CB bands or with excessive power have received five-figure fines over the past couple of years as enforcement activity has heated up. Ask yourself if you need the potential hassle just to talk on the radio. If you want to run higher power or talk to foreign stations, that's exactly what the amateur radio service is for!

Chapter 6

Communicating in Emergencies

* *

In This Chapter

▶ Using radios for emergencies

▶ Preparing for emergency communications

▶ Saying the right thing

▶ Being part of an organized disaster response team

▶ Getting necessary training

* *

*O*ne of the main reasons many people buy two-way radios is to communicate in emergencies. Radios are great for this purpose! Two-way radios don't require the assistance of any kind of telephone or data networks, relay towers, or cables — all of which are likely to be out of service or highly congested in the event of a disaster. Obviously, in an emergency you would want to talk with family members and close friends. But with whom would you *actually* be able to communicate — and *how?* This chapter introduces a number of important basics of *emcomm* (or *emergency comm*unication). To go beyond this chapter, I recommend Dave Ingram's fine book, *Emergency Survival Communications* (Universal Electronics).

Matching Radios and Emergencies

In order to effectively communicate during an unforeseen emergency, you have to make some decisions now about what you hope to accomplish. Knowing your communications goals is essential if you are going to match the right radio to your needs. As you've probably seen already just flipping through this book, you have several different options to choose from.

Emergencies can take a number of forms and occur at any time. Some circumstances when a radio could come in handy include

✔ A major storm, such as a hurricane or a tornado

✔ An earthquake

- A wildfire
- An accident on the road, train rails, or in the sky
- An accident while camping, kayaking, hiking, or hunting

Choosing the right radio is much easier if you think like a reporter covering a story — ask yourself *who* you want to talk to, *what* you plan to say (and in what format), *why* this format makes the most sense, *when* you will use this form of communication, and *how* you plan to do it.

Deciding who you plan to talk to

In the event of an emergency, *who* will you be trying to talk to? The answer is probably not just a single person, but certainly must be limited — you won't be able to talk to anyone and everyone you know. The best way to determine who you want to contact is to assign communications categories. Prioritize your list in order of urgency. Start with the health and safety of you and your family. Even emergency workers are trained to be sure their home and family are secure before responding to other parties. After that, decide what's most important to you and arrange the list accordingly.

You will need to contact more than one group, so start with the easiest groups to define — your family members. Will you be trying to contact only immediate family members? How about a distant relative? Which of these people do you expect to operate or have access to a radio?

It's a good idea to have an out-of-state contact so that others can contact that person about you. Under emergency conditions, authorities often restrict or block *health-and-welfare* radio queries coming into the affected area. However, personal information can flow out fairly easily, so getting your status to the distant contact allows that person to inform others.

When you have hashed out which of your family members you will contact, you should widen your perspective. Here are some groups to consider:

- **Neighbors and members of the local community:** Depending on where you live, your nearest neighbor may be a long way off and unreachable by phone. Even in densely populated urban areas, having some other option besides face-to-face contact is very helpful. This is the time to think about which of the people around you have special needs (so you can help them) or special equipment (so they can help you).

    Keep thinking in ever-widening circles.

✔ **A contact for your local public safety agency or emergency response organization:** CB and marine radio both have dedicated emergency channels. Who would be monitoring those channels?

✔ **A coordinator for church, social, service, or recreational organization responses:** Parents will want to know what's going on at school.

Make a list of all these contacts and start recording all of their *coordinates* — physical address, phone numbers (fixed and mobile), and whether they have radio capabilities. A computer spreadsheet is an excellent and inexpensive way to collect and organize data.

Prioritizing the what and why

When you decide *who* you want to contact, you also have to consider *why* these contacts make your final cut and *what* information you'll want to exchange. The reasons may be self-evident — you want to contact your mom because she worries, your neighbor because he lives alone and uses a wheel-chair, and so on.

Be sure to write down the nature of the exchange you are likely to have with each of your contacts, as well as the specific information you need to exchange (the *what,* in journalistic terms). For example, you may need to give directions to your house or a description of a medical condition.

Don't rely on memory for this information. Even the best-trained individuals can go blank under stressful circumstances. Write everything down so that all you have to do is read it out. Having the information in written form and clearly labeled will pay big dividends if you're incapacitated and someone else is communicating on your behalf.

You may also need to provide identifying information or a membership number in order to give and receive information. For example, schools often require a code word to give out or accept student information. Write that down, too!

Firming up how and when you'll communicate

The final step in making the list is to add information about the means of communications for each contact. Your family members may have a mix of

FRS/GMRS and CB radios. A business may monitor a CB channel. Boaters rely on Marine VHF. You need that information in your list just like phone numbers.

To take care of the *how* for each contact, list the type of radio and channel or frequency for each contact. For example, add to your list something like the following: *Family uses FRS/GMRS Channel 8* or *The Wilsons monitor CB Channel 9 from their RV.*

The final step is to identify *when* you will try to make contact. If you know the contact's schedule of availability, include that information so you attempt contact at the right times. If you don't know, this would be a good time to find out! For example, your family might agree to try meeting on the air at the top of every hour. Use your local time to avoid confusion on your part.

Here are some suggestions:

✔ If you are trying to contact family or friends near your home, standardize on a channel of any convenient radio service and use a repeating time, such as *top of the hour* or *half-past the hour.* Don't get complicated; keep the number of channels and times to a minimum and be sure to have an alternate.

✔ If you are away from home, use the ARRL Wilderness Protocol (see Chapter 4) or a twice-a-day contact period. Be sure to take a longer-range radio, such as a CB, that can get through to other people from where you expect to be.

✔ If you're trying to get word out of an affected area, tune to the emergency channels and listen for disaster relief organizations, such as REACT or a public safety agency. Ask to have a message relayed for you and have it ready and written down. Keep the total message length to 25 words or less. Include the complete contact information for you and the addressee.

Filling in the blanks

When you've completed the list, you'll have a lot more questions than when you started! Now is the time to start working on that missing information. During an emergency is no time to start asking how to get in touch with a boat repair shop or local medical clinic. Print out several copies of the list (with the date on which the information was current) and keep them in the places you're most likely to be.

Here's some information about what you should include in your list:

✔ **Only the most succinct facts:** Sifting through a short novella (even if the plot *is* really good) isn't what you want to be doing in an emergency. Keep it short.

The information on your list is often confidential and personal. If you plan on traveling with the information, remove all but the most necessary items. If you lose the information or it's stolen, protect yourself from identity theft by notifying financial and medical institutions.

✔ **Sources of information that you will need:** For example, if your county or town has broadcast stations that are designated as official sources of information in an emergency (check their Web sites), note them. National Oceanic and Atmospheric Administration (NOAA) weather channels are excellent sources of weather information, particularly if you have a receiver with SAME (Specific Area Message Encoding) technology. Check out the National Weather Service's NOAA Weather Radio site for more information (www.nws.noaa.gov/nwr/index.html). Include the frequencies used by local public safety agencies (see Chapters 10 and 11), to get real-time information on local activities.

Being Ready

Being successful in communicating during emergencies only occurs if you are ready to be successful. That sounds trite, but attempts at using your radios will fail if you don't know how to use the equipment or the equipment isn't ready. When the equipment and knowledge are in place, you only have to keep them fresh and freshen yourself up with a little practice now and then. Here's some advice on keeping your radio equipment and yourself ready for an emergency:

✔ **Know how to operate your radios:** The moment your adrenaline begins pumping is not the time to ask, "How do I access the memory channels?" Keep the manual with the radio or in a *go kit* (I cover go kits in the sidebar elsewhere in this chapter). For those hard-to-remember functions, make up a cheat sheet and attach it to the radio.

✔ **Keep your batteries ready:** Even rechargeable batteries wear out over time. Follow the manufacturer's directions on battery maintenance. Rotate your battery packs in and out of service. Keep spare alkaline batteries in a refrigerator (or at least keep them dry) to maximize their shelf life.

✓ **Test the radios regularly:** Do more than just turn them on and listen now and then. Make contact with someone! That will exercise the radio, the antenna, and any connecting cables. If you have an external or mobile antenna, inspect it once every few months to be sure nothing is broken or damaged.

✓ **Keep your information current:** Every so often, you should go through the information list, updating addresses and locations. Verify that the channels, phone numbers, and other information you've listed are still correct.

Making and Responding to Calls for Help

When you are in an emergency, the most important thing you can do, especially if you have been injured, is collect yourself, regulate your breathing, and try to maintain a calm, *slow* speaking voice. If you're receiving an emergency call, try to be a soothing voice to the person you're speaking with, and, if you have paper and a pen nearby, write down everything you hear. The following sections give more advice.

Making a call for help

If you need immediate emergency assistance, regardless of your location or situation, the appropriate voice signal is, "Mayday" (just as in the movies). Repeat the call over and over until you get an answer or for as long as possible. If you are attempting to interrupt an ongoing conversion, repeating the word *break* twice ("Break, break!") is often used as an emergency signal.

Call channels to which others are likely to be listening, such as an emergency channel. Have these channels stored in your radios' memories, if possible. If you do not receive an immediate response when you use an emergency channel, repeat the call and include the following information:

✓ The location or address: "I'm near the Rushing River downstream from the Highway 9 bridge."

✓ The nature of the problem: "I'm lost."

✓ What type of assistance is needed, such as medical or transportation aid: "I hurt my leg and can't walk."

✓ Any other information that might be helpful: "I can see Mt. Baldy to the West."

This is smart emergency procedure; if someone can hear you but you can't hear him or her response, your information will still be relayed to the authorities.

If there is an immediate threat to life or property, anyone, licensed or not, can use whatever radio equipment is available to call for help on any frequency or channel.

Receiving a call for help

If you hear a distress signal, immediately find something to record information (a pad of paper and pen or pencil); note the time and channel of the call, and then respond by letting the caller know that you can hear him or her.

For example, on CB, you might say, "Clean Jean, this is Rubber Ducky. I hear your distress call. What is your situation?"

In order to allow the proper authorities to render assistance as quickly as possible, record all the information you're given by the caller, even if the station doesn't respond to you. When you feel that you have all the information the station can or will provide, call the appropriate public agency or public emergency number, such as 911. Explain that you have received a distress call by radio and give the channel or frequency. The dispatcher will either begin asking you for information or transfer you to a more appropriate agency.

After notifying the agency, return to the radio and let the caller know who you have contacted and any information you are instructed to relay.

Do not embellish the information in any way! Attempt to be 100-percent accurate in relaying information. You may be asked to act as a go-between until representatives of the agency or authority are in direct contact with the station in distress and release you from further communications.

Disaster Response

Because disasters affect large numbers of people, emergency communications is usually conducted by organized groups, both government and private. There are many such organizations and they actively seek volunteers to support their operations. Some function as independent groups; others are auxiliary to government agencies. If you're interested in assisting your community, you can put your radio skills to good use by joining the local disaster response organization. The following list includes a few of the better-known groups, but there are many more.

- **REACT (Radio Emergency Associated Communications Teams):** REACT promotes the use of radio to assist in emergencies. It is organized on a regional basis and provides training opportunities, working with public safety agencies (www.reactintl.org).

- **CERT (Community Emergency Response Teams):** CERT teams are organized by a local public safety agency, such as a fire or police department to supplement the agencies' resources in disasters. Communications is an important part of CERT functions, but members receive training in first aid and rescue techniques, as well (www.citizencorps.gov/programs/cert.shtm).

- **American Red Cross:** Providing disaster relief requires an extensive communications presence. There are a number of ways in which you can use your radio skills to assist the Red Cross (www.redcross.org). Outside the U.S., the International Committee of the Red Cross (icrc.org) is the organization to contact.

- **ARES (Amateur Radio Emergency Service) and RACES (Radio Amateur Civil Emergency Services):** These services use ham radio to support public safety and federal emergency agencies, respectively. As I mention in Chapter 9, even entry-level hams can play a valuable role in these groups (www.arrl.org/FandES/field/pscm/sec1-ch1.html and www.races.net).

- **National Association for Search and Rescue:** If you are physically fit and like the outdoors, search and rescue organizations are waiting for you! These teams often support the local county Sheriff's Department (www.nasar.org).

- **SKYWARN:** Not strictly a radio group, SKYWARN was formed to provide up-to-the-minute information to the National Weather Service on weather conditions. Information is provided directly from the field by radio or phone (www.skywarn.net).

Not all these groups are active in every region and there are many other groups that I haven't listed. In fact, you may be part of an organization right now that has its own emcomm team or plan. The best way to find out what groups are active in your area is to contact your local police or fire department or the Red Cross chapter in your area. These organizations are already working with groups and would be happy to assist you in connecting with them.

Setting up a go kit

Whether or not you intend to join a formal emcomm group, I recommend having a *go kit* for emergencies of any sort. You have to create a go kit appropriate for your needs, of course, but here is some guidance on what to consider. The following is adapted from the Spring 2003 *Radiogram,* the Western Washington Emergency Services Team's quarterly newsletter:

In its most basic form, the go kit is a group of items, prepared in advance, kept in a common carrying case of some sort, that allows the emergency responder to have equipment and supplies that will allow them to function independently for a specified period of time. Go kits may be designed for a 24-, 48-, 72-, 96-hour, or longer response. The benefit in having this kit previously prepared so that it is put together in a time when you are not actually trying to get out the door. Under more relaxed conditions, you are much less likely to forget important and/or useful items.

If you are putting together a 48-hour kit, you would likely include the following items:

✔ **Food:** You should include food and water. This is especially important if you have special dietary needs. You may or may not have food available. Your rations should include foods that do not require refrigeration. You should consider military surplus *MREs.* You may also consider trail mixes, protein bars, dried fruits, even canned foods. Many missions would not be conducive to heavy or bulky supplies however, so be sure to consider what your needs are likely to be.

✔ **Clothing:** Update your go kit with appropriate clothing for the season. Make sure your clothing is comfortable as well as expendable. But most important, make sure that your clothing fits your needs. Not only is it to be comfortable (as near as possible), but it also needs to offer you some level of protection in emergency conditions.

✔ **Radios:** Make sure you are bringing everything you may need, including the radio(s), microphone, antenna, spare batteries and/or power supply, and adapters. You may also want to bring some sort of broadcast receiver that will allow you to keep informed of news from the outside. This may mean several back-up battery packs, gel-cells, or recharging units. A back-up antenna may be useful if it is lightweight and relatively easy to set up.

You should also have a prepared list of operating channels or frequencies, a personal phone book with numbers of friends, family, and other useful numbers. Remember, if you think these things out in advance, you will be far better prepared than if you are running through your home, throwing together a kit as you head out the door. Be sure to have a checklist of items you will need and use it!

Practice Makes Perfect

Regardless of whether you join a formal emcomm group, there is no substitute for regular radio practice, whether just in the course of regular recreational activities or an organized event. The examples in Chapter 1 are great starting points.

The more you participate in organized communications activities, the more skilled you'll become. Should you ever participate in an actual emergency or disaster response, those initial butterflies will quickly dissipate and you'll feel right at home using your radio.

Chapter 7

Workaday Wireless: Business Radio Services

In This Chapter

▶ Using a business radio

▶ Determining which business radio is right for you

▶ Getting a license

▶ Working with radio professionals

*B*usinesses use radio technology in many ways, and you have several options to choose from. In this chapter, I cover the different types of radio services that are available and present to you the commercial radio options that are offered from professional radio companies. Most business radio services require a license to use, so I touch on the necessary paperwork you need to fill out to stay on the right side of the FCC. I wrap up the chapter with some information on operating a business-type radio.

Choosing the Right Business Radio Service

If you read Chapter 4 on FRS/GMRS radios and aren't convinced that they meet your needs, I'm guessing that you need a more comprehensive radio solution. You fall into this category if you have a lot of employees on site; maybe you have delivery or sales personnel traveling to customer sites.

Non-U.S. readers may find that most countries have equivalent services under their local licensing authorities; for example, in Canada visit Industry Canada (sd.ic.gc.ca); visit the Office of Communications (www.ofcom.org.uk) in the U.K.

The following two sections offer information about two radio services that are more robust in their offerings than good old FRS/GMRS.

Multi-Use Radio Service: MURS

Technically a part of the FCC's Personal Radio Service, *Multi-Use Radio Service* (MURS) is the most capable of the unlicensed two-way radio options in the United States. Here are the specifics:

- ✓ **Frequency:** MURS operates on four channels near 154 MHz.

- ✓ **Power and antennas:** Radios are limited to 2 watts of output power and antenna height can be no greater than 60 feet.

- ✓ **Range:** Because repeaters are not permitted, range is limited to about 20 miles at best.

MURS is very popular with individuals and small businesses. MURS makes sense for smaller organizations with up to a dozen or so people using hand-held radios. You can use an external antenna to extend the range to a few miles, but without repeaters, individual users will find it difficult to contact each other if they're more than a mile apart. MURS is particularly useful on temporary job sites. MURS channels are shared, so you can't always count on a clear channel, but the cost of the radios is low. MURS is described in detail on the Public Service Radio Group Web site at `www.provide.net/~prsg/murshome.htm` and the FCC's MURS home page is `wireless.fcc.gov/services/personal/murs`.

Private Land Mobile Radio Services

If MURS is too limited, step up to a fully licensed radio system within one of the *Private Land Mobile Radio* (PLMR) services. Also referred to as the Business Radio Service, or BRS, PLMR radio systems include everything from simple radio-to-radio (one-to-one) operations for use around a mall or warehouse to complex trunked systems (which I describe in Part III) that cover huge areas and thousands of radios. You can operate on channels shared between many users or, if you decide to spend the money, build a radio system exclusively for your own business to use.

With PLMR, you'll find yourself in the Industrial/Business Radio pool.

Here are some other details about PLMR:

- ✔ **Frequency:** PLMR operates on sets of channels with frequencies that range from VHF Low Band to 1400 MHz.

- ✔ **Power:** Radios range from handheld devices with a few watts of output to power mobiles and base stations of 100 watts or more.

- ✔ **Range:** Up to regional coverage is available with a repeater or trunked system.

The FCC's Web site on the Industrial/Business Radio pool of PLMR is `wireless.fcc.gov/services/ind&bus`. The variety of different organizations using PLMR radios is staggering, so it's difficult to summarize. I suggest that you evaluate GMRS and MURS radios first. If those systems can't handle your needs, PLMR radio is for you and you need to get professional assistance as described in the next section.

Using a Professional Radio Service Provider

Because you're just finding out about using two-way radios, I strongly recommend that you find a local vendor to guide you through the process of selecting radios. Why? Cost. If I was an Internet expert and I was advising a group of beginners about setting up a home network, I would say the same thing. You *could* act as your own Internet Service Provider by arranging for the high-speed connection, installing network equipment, and then maintaining it all, but that usually isn't cost-effective. Why not turn over the details to a third-party specialist? Radio systems of any significant size are no different.

Unless your business has a compelling reason to have its own radio shop, it is customary to use a local radio system expert. This section reviews the most common arrangements.

Start by assessing your communications needs:

- ✔ How many people will need to use the system?

- ✔ Will they be in one building or spread out?

- ✔ Do they need to talk to each other, or just to a central office?

- ✔ Will they be in vehicles, walking around, or at fixed locations?

- ✔ Will your needs grow?

Consult your trade or business organizations for information on how radio systems are used by others like you. Then contact a couple of local vendors. You can usually find listings in business directories under the heading "Radio Communication Equipment and Systems." Look for a company that advertises "radio systems" and offers the following services:

- **Requirements assessment:** You need to talk to someone who can help you determine your real radio needs so that you don't buy too big or too little a radio system.

- **Licensing assistance:** Most shops have all the necessary paperwork on hand and are experts in the filing process.

- **Radio selection:** Look for a vendor that distributes at least two major radio brands.

- **Sales, lease, or rental services:** Commonly, these businesses offer packages of equipment or service.

- **Installation, programming, and maintenance:** Although the initial installation is the most critical, you'll need a service plan that is flexible enough for your needs.

Radio system terms

After you select a radio system provider, you are thrown into world filled with lingo. If you feel as though everyone else is speaking a different language, don't worry — educate yourself. To understand what kind of system the vendor is designing for you, know the following terms in the following contexts:

- **Analog systems/digital systems:** Analog systems are systems whose signals can be received and understood directly. Digital systems convert voices into a digital format before transmission. Analog is the most popular and least expensive option.

- **Encrypted signals/scrambled signals:** Encrypted signals are altered digital signals that can't be accessed by unauthorized parties. Systems can be configured to encrypt signals by converting them to digital codes before transmission. Scrambling is an analog process.

- **Multicast/simulcast:** A multicast system is one in which transmitters use a set of different frequencies and the receivers tune to the channel with best reception. A simulcast system provides wider coverage with multiple transmitters, all of which broadcast the same signal simultaneously on a common frequency.

✔ **Multisite/single-site:** A multisite system extends coverage beyond single-site systems by using two or more repeaters or transmitters linked together. A single-site system has just one base station or repeater, covering an area centered on that transmitter. Operation can be dispatcher controlled or one to one.

✔ **Shared or subscribed system:** A system in which a vendor rents shared space to small businesses that need wide coverage but can't afford to purchase a repeater.

✔ **Voting:** The process of selecting the best signal from multiple satellite receivers to be transferred to the repeater or dispatcher. This occurs when multiple satellite receivers are added to increase the receive range of the system. A comparator circuit selects the best signal from the receivers to be transferred to the repeater or dispatcher.

✔ **Wideband/narrowband:** *Band* refers to the range of frequencies taken up by a signal. Narrowband equipment occupies 12.5-kHz channels, as compared to the older wideband technology, which uses 25-kHz channels. Systems should be all wideband or all narrowband.

A subscription to *MobileRadio Technology* (MRT) magazine (`mrtmag.com`) is a good way to keep up with technology and business radio industry development. The magazine publishes a yearly buyer's guide. Also, every issue contains ads from vendors.

Staying in touch with dot and star channels even when you're itinerant

Commonly, job crews or personnel work at temporary sites away from a home office or headquarters. Of course, you still want to stay in contact on the job site by radio if possible. You could find a local radio vendor and set up a short-term subscription to its system. Another (better) solution is to use the channels designated for *itinerant* users. This solution is a much less expensive option for mobile business users such as entertainment companies, construction crews, and even circuses.

The special channels are part of the PLMR service. They are known as the *dot* or *star* channels because radios that operate on these channels use colored dots or stars to label the channel select switch. This makes managing radio use easy — all you have to do is say, "Let's use red dot," and all users will tune their radios to the same channel at 151.625 MHz. (You can find a comprehensive list of business channels at `www.freqofnature.com/frequencies/common/business.html`.) Repeaters are not used on these channels, so communications is strictly short range (less than a mile).

Business radios for the dot and star channels are sold over the counter without proof of license. If you decide to use these channels for your business, you are still required to file the required FCC paperwork and obtain a proper license.

Licensing your business radio

The simplest way to get licensed is to apply, using your radio vendor. Vendors know the ropes and the most efficient way to apply. If you've never applied for a business radio license before, let the vendor do it for you.

To make the process go smoothly, review the licensing process as described in Chapter 3. You can get a good description applying for a business licensing on the FCC's Wireless Telecommunications Bureau Web site at `wireless.fcc.gov/services/ind&bus/licensing`.

Here are the general steps:

1. **Start by getting your FRN (see Chapter 3) if you don't already have one.**

   This is an easy process and saves your vendor a step.

2. **Collect all the necessary information to prepare your preliminary application.**

   These items are listed at `wireless.fcc.gov/services/ind&bus/licensing/prepchecklist.html`. Some items are fairly technical, so the radio vendor will supply them.

3. **The vendor submits your application via one of the FCC frequency coordinators.**

   Coordinators help minimize interference by managing the number of users on a channel.

4. **You may start building the system and operating ten days after the frequency coordinator accepts the application.**

   Or you can begin building the system when your license is granted (your vendor will know when this occurs).

Your license covers not only the frequency, but also location of the antenna, how much power the transmitter puts out, and so on. Any time you want to change any of these, you are required to refile your licensing information. This allows the frequency coordinators to minimize interference that might occur from a station raising its antenna or transmitter power.

Operating Your Business Radio

Although the licensing process can be complicated, business radios are typically quite simple. After all, the goal is to communicate with a minimum of distraction. To that end, business radios typically have the fewest controls needed to get the job done. The Icom F21BR radio, shown in Figure 7-1, is a typical example. Mobile and base station radios have the same controls:

- ✔ **Volume:** Often combined with Power On/Off as a rotary switch. Some radios have a power button the user must press and hold to activate.

- ✔ **PTT:** Push to Talk keys the transmitter. See Chapter 4 for a photo of proper microphone use.

- ✔ **Antenna:** Most business radios have detachable antennas. Some antennas unscrew; others use a *bayonet* connector that attaches with a press-and-turn action.

- ✔ **Channel Select:** The simplest of radios may have just one or two channels, selected by a labeled switch. Multichannel radios have a knob that selects from up to 16 channels. A priority channel switch that overrides the knob setting may also be present.

- ✔ **Monitor:** Pressing this switch opens the squelch circuit so that the user can hear weak signals. For a complete discussion of squelch and monitor functions, see Chapter 4.

- ✔ **Wide/Narrow:** This button allows users to switch between wideband and narrowband operation.

Operating rules

A shared or dispatched system has a set of operating rules and procedures to control the flow of information. The more complex the system and the greater the number of users, the more tightly *radio discipline* must be controlled. Check with the FCC rules for information about identification requirements or other regulatory rules.

If you have a small system with just a few regular users, you can be very informal — radio on a first-name basis. The keys to making business radio work in any setting (regardless of the size of your group or the complexity of your system) are enumerated in the following list:

- ✔ Keep transmissions short

- ✔ Discourage nonwork chatter

- ✔ Speak clearly

- ✔ Use a standard set of abbreviations or codes

- ✔ Keep confidential business information off the air

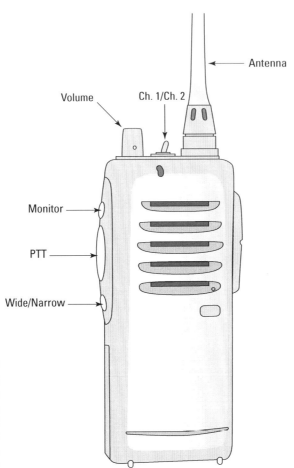

Figure 7-1:
The Icom
F21BR is a
typical
business
radio.

Radios for *trunked systems* have more controls and usually have an alphanumeric display that allows the user to change *talk groups* and perform other system functions. Trunked systems are described in more detail in Part III.

Chapter 8

Ladies and Gentlemen, Ships at Sea: Marine Radio

In This Chapter

▶ Using marine radio around the harbor and on open water

▶ Selecting and operating marine radios

▶ Evaluating satellite radio and e-mail systems

▶ Getting a marine radio license

*B*oating and radio have been partners for nearly a century. Mariners took to wireless like ducks to water, instantly ending their isolation at sea. Today, all boaters need radios; many novices make a shiny, new radio their first marine electronics purchase. Mobile phone service is often unreliable on the water and land-based data networks are out of commission as well; the marine radio functions as both a navigational assistant and a portable emergency lifeline. Marine radio's greatest boon to the maritime community has been improved safety.

Marine radio is a rather broad field. To follow up or explore the information in this chapter, I recommend Chuck Husick's discussion of marine radio hosted by BoatU.S. at www.boatus.com/husick/default.asp. The U.S. Coast Guard also provides a comprehensive overview of marine navigation systems at www.navcen.uscg.gov/marcomms/default.htm.

This chapter gives you a fish-eye view of VHF and HF marine radio — with information about both the controls and operation. We also skim the surface of keeping in touch with e-mail while boating. Because radio performance depends heavily on installation quality, I augment the general information in Part V with tips specific to marine installation. Licenses are required for some marine radio operation, so this chapter shows you how to pick up your ticket (that's radio jargon for getting licensed).

Introducing VHF Harbor and Waterway Radio

Most boaters, whether they use power motors or sail, cruise within sight of land, even in salt water. This makes radio communications with other boats and shore-based stations fairly straightforward. Rather than tie up long-distance frequencies with short-range chatter, the VHF marine band was created for harbor and waterway use. Because these are VHF channels without repeaters, the range of communications is from 5 to 10 miles.

Located on frequencies between 156 and 162 MHz, the marine band takes up 48 regular channels (mostly *simplex*) for voice communications and a pair of channels for the Automatic Identification System (discussed later in this chapter). There is some variation in channel use around the world. A complete list of all international marine VHF channels is available at www.marine-electronics.net/techarticle/vhf/intvhf.htm.

Unlike CB and FRS/GMRS radios, with which anyone is allowed to use just about any channel, all the marine VHF channels have specific uses, as shown in the complete table at www.navcen.uscg.gov/marcomms/vhf.htm. Table 8-1 lists the channels used by noncommercial, recreational boaters in the U.S.

Table 8-1	Noncommercial U.S. Marine VHF Channels		
Channel Number	*Ship Transmit (MHz)*	*Ship Receive (MHz)*	*Use*
6	156.300	156.300	Intership Safety.
9	156.450	156.450	Commercial and non-commercial boater calling.
13	156.650	156.650	Intership navigation safety (bridge to bridge). Ships >20m length maintain a listening watch on this channel in U.S. waters.
16	156.800	156.800	International distress, safety, and calling. Nearly all ships, USCG, and most coast stations maintain a listening watch on this channel.
22A	157.100	157.100	Coast Guard liaison and maritime safety Information broadcasts. Broadcasts announced on Channel 16.

Channel Number	Ship Transmit (MHz)	Ship Receive (MHz)	Use
24	157.200	161.800	Public correspondence (marine operator).
25	157.250	161.850	Public correspondence (marine operator).
26	157.300	161.900	Public correspondence (marine operator).
27	157.350	161.950	Public correspondence (marine operator).
28	157.400	162.000	Public correspondence (marine operator).
68	156.425	156.425	Noncommercial.
69	156.475	156.475	Noncommercial.
70	156.525	156.525	Digital selective calling (voice communications not allowed).
71	156.575	156.575	Noncommercial.
72	156.625	156.625	Noncommercial (intership only).
78A	156.925	156.925	Noncommercial.
79A	156.975	156.975	Commercial. Non-commercial in Great Lakes only.
80A	157.025	157.025	Commercial. Non-commercial in Great Lakes only.
84	157.225	161.825	Public correspondence (marine operator).
85	157.275	161.875	Public correspondence (marine operator).
86	157.325	161.925	Public correspondence (marine operator).
AIS 1	161.975	161.975	Automatic Identification System (AIS).
AIS 2	162.025	162.025	Automatic Identification System (AIS)

Here is a little basic advice for using the various channels:

✔ **International calling, distress, and calling channels:** Channel 16 is the international calling and distress channel and Channel 9 is for calling other vessels. Don't use either channel for extended conversations. They're just for you to make your call and establish contact. Unless it's an emergency, you should immediately move to one of the non-commercial channels.

✔ **Commercial and government channels:** You may not use commercial or government channels for day-to-day conversation, even when they seem clear. At the very least, you'll likely be rebuked by the proper users; you may even receive a notice of violation or fine. Descriptions of the intended uses are published at www.marinewaypoints.com/learn/vhf.shtml.

✔ **News and information:** Channels 6, 13, and 22 are all good sources of information during bad weather or in an emergency. You shouldn't transmit on them, except for Channel 13 to coordinate navigation with another boat or talk to a land facility such as a drawbridge operator.

✔ **Contacting shore operators:** Marine operator channels (24 through 28 and 84 through 86) are for contacting shore operators that will place phone calls for you (for a fee). Conducting a phone call over the radio is called a *phone patch*. Not all the marine operator channels are staffed in all areas. To find out whether an operator is available (and on what channel), call for a marine operator on Channel 16.

Knowing your marine VHF radio controls

Just as in the other chapters on specialized radio services, here's a guide to the typical marine VHF radio interface. Because marine VHF communications use FM signals, the controls and functions, shown in Figure 8-1, are very similar to those on FRS/GMRS (Chapter 4) and business radios (Chapter 7).

Squelch, volume, power, PTT, and channel select controls and indicators are all described in detail in Chapter 4. Here's what's different for marine VHF radios:

✔ **PA:** The public address speaker is known onboard as a *hailing speaker* and operates similarly to the PA of a CB radio (see Chapter 5). Keep the hailing speaker away from the microphone to avoid the squeal of feedback.

✔ **Transmit Power:** Pressing this control toggles between high and low transmit power. An indicator on the display shows the power level. If you're having trouble making contact, be sure you're not using a low output power.

✔ **Weather Alert:** The weather alert function listens for a coded or tone alert from the National Oceanic and Atmospheric Administration's weather channels. The NOAA's tone alerts indicate severe weather.

✔ **Priority Channel:** Press this button to switch the radio to the distress/calling channel. The Oceanus can continually monitor both Channel 16 and 9 and whatever channel you're operating on. Most marine radios have a similar feature that listens to these channels, even while you're performing regular radio communications on a different channel. A Priority Watch feature automatically switches the radio to Channels 16 or 9 if activity is heard there.

✔ **Distress Call:** This may be the most important switch on the radio. Pressing it activates a distress call function, which attracts the immediate attention of the Coast Guard and other boaters. This button is often marked in red and covered with a protective shield to prevent accidental activation. Obviously, this is not a button you should play with unless you're in immediate danger.

You should apply the same guidelines in Chapters 4 and 5 for using a microphone so that your voice is transmitted clearly. Be particularly aware of wind noise — speak in a sheltered area or use your hand or body to shield the microphone from the wind.

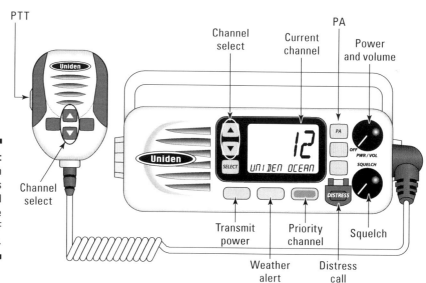

Figure 8-1: The Uniden Oceanus is a typical midrange VHF transceiver.

Choosing a marine VHF radio

A few considerations will dominate your choice of radios. They are driven by your boating habits and where you plan on installing and using the radio:

✔ **Handheld radios versus permanent-mount radios:** Handheld portable units are great for kayakers, canoeists, and the occasional boater who doesn't want a lot of electronics. A handheld radio also doesn't rely on the boat's battery to work. The drawback of handheld radios is primarily their low power. Handheld receiver performance is usually not as good as with larger permanently installed radios.

You may want to purchase a handheld radio to use in a dinghy or on shore as a backup for your permanent-mount radio.

✔ **Receiver selectivity and sensitivity:** You want your radio to receive only the desired channel in the presence of strong signals on nearby channels. For example, if you expect to use the radio in a busy harbor or waterway, selectivity is important. Compare the *adjacent channel selectivity* and *rejection* specifications — figures of 60 dB (decibels) or higher are better. Receiver sensitivity (the ability to hear weak signals) is roughly equivalent on almost all radios.

✔ **Transmit power:** An output power of 25 watts (permanently installed) or 5 watts (handheld) is adequate. Look for a radio that can reduce power to 1 to 5 watts in a crowd. Higher power is needed if you anticipate boating in isolated areas.

✔ **Water resistance:** Manufacturers may specify that their radios meet a standard for water resistance. For example, ASTM B117 is a standard that is used to assess the ability to withstand saltwater spray. Several manufacturers refer to the JIS-7 waterproofing standard. Look for information in the radio specifications that indicates that the radio meets one or more of these standards. If you plan on mounting the radio in an exposed position, the higher ratings are worthwhile.

✔ **Weather alert:** One of the most common extras is a weather alert feature. If you will be traveling in a region with severe weather, this is a desirable feature, although it's not strictly necessary.

✔ **NEMA interface:** A NEMA (National Electrical Manufacturers Association) data interface is required if you plan on integrating a GPS (Global Positioning System) or LORAN (*LO*ng *RA*nge *N*avigation) receiver into your communications system for advanced position reporting. However, if you're not going to be doing a lot of long-term, open-water navigation, it's not necessary.

Choosing an antenna for your marine VHF radio

When it comes to antennas, here's what you need to know. Most marine VHF antennas are *whips* (flexible, slender rods, often enclosed in protective fiberglass housing). Unlike whip antennas designed for vehicle mounting, their marine cousins do not require a ground plane or radial wires. (Most boats have a nonconductive hull or deck that doesn't provide a very good ground.) Marine whip antennas don't require a connection to a ground and are designed to be bolted or clamped to any supporting structure. As long as they are mounted in the clear, they work reasonably well.

You may run across the following terms as you shop around for an antenna:

- **Half-wave whip:** The metal rod or wire is half a wavelength long, which is about 36 inches for marine VHF channels.
- **Collinear:** Longer than the half-wave whip, the collinear is a series of half-wave antenna sections connected together inside a protective tube to create a more powerful signal broadside to the antenna. A collinear antenna can double the range of your signal, which is important if you regularly travel outside harbors or away from regularly traveled waters.

Introducing Marine Radio's Advanced Features

With a growing number of vessels plying the waters over wider ranges, modern technology was needed to organize and accelerate marine communications. By incorporating digital technology, making contact and reporting information have been greatly improved.

Digital Selective Calling (DSC)

To contact a specific ship by voice, use the calling channel and hope the skipper is listening. Digital Selective Calling (DSC), developed by the U.S. Coast Guard and International Radio Consultative Committee (CCIR) in the late 1980s, takes that idea to the next level by assigning every ship a unique identifier called a Marine Mobile Service Identity (MMSI). The MMSI is a unique nine-digit number stored inside your radio. You can elect to obtain an MMSI when you get your marine radio license. Not all radios have DSC capabilities, so check the specifications before you buy. The casual boater probably doesn't need DSC, but if you're on the water a lot or need regular communications with particular boats or facilities, it's a good feature.

The *selective* part of DSC refers to the ability to contact a particular vessel by using MMSI identifiers.

To make a call, using your radio's DSC capabilities, you must know the MMSI of the radio of the other vessel. Enter the MMSI and select the voice channel you want to use after contact is established. (Some radios can store a directory of MMSI data.) The radio does the rest on DSC Channel 70. If the other vessel responds, the contact channel information is passed and the radio operator on the other ship is alerted. The users never actually talk directly on Channel 70.

There are two types of DSC — class A and class D. Here's a breakdown of their offerings, and how they affect the cost of your radio:

✔ Class A radios require a separate receiver that's tuned to Channel 70 at all times. Class A DSC radio offers more reliable connections but adds substantially to the cost of the radio.

✔ Class D radios are only tuned to Channel 70 under operator control. They're less expensive but do the job just fine.

Feeling less sea sick with GMDSS

For many years, listening for and responding to distress calls from mariners relied on the U.S. Coast Guard's 24-hour watch on known emergency channels and frequencies. Although this system worked reasonably well, it still required a lot of effort to staff and to make the distress call in the first place. In the late 1970s, the International Maritime Organization (IMO) began developing a new system of maritime distress and safety communications.

In 1979, a group of experts drafted the International Convention on Maritime Search and Rescue, which identified the need for a global search and rescue (SAR) plan. The Global Maritime Distress and Safety System (GMDSS) is the implementation of that plan.

In the old days, SAR was primarily a ship-to-ship operation. The GMDSS uses both satellite and terrestrial radio services to create a ship-to-shore system that uses *rescue coordination centers* located at military bases and coast guard facilities. GMDSS automatically handles distress alerts and identifies the location of the emergency, even when the radio operator doesn't have time to send a complete SOS or Mayday signal. GMDSS equipment supports distress and safety alerts (including providing position information), search and rescue coordination, homing, safety broadcasts, and general communications.

DSC radios that have radiolocation gear attached (such as GPS or LORAN) can send a DSC distress call on Channel 70 that includes the location of your boat, as well. Your radio must have the NEMA data interface to read the data from the radionavigation gear.

Automatic Identification System (AIS)

Automatic Identification System (AIS) communications occur on the two channels just above 162 MHz; the channels are called AIS1 and AIS2, and they're used by suitably equipped radios to broadcast information, such as a vessel's MMSI, course, direction, speed, and so on. If you're suitably equipped with an AIS position monitor, you can display a radar-like map that shows the positions of all other ships whose AIS broadcasts you receive. AIS is usually used by larger vessels and by shore monitoring stations.

Saltwater Communications: HF Marine Radio

When traveling on the open ocean a long-distance means of communication is required. This is where HF (high frequency) radio excels. Requiring no repeaters or satellites, HF signals are bounced back to earth by the ionosphere, traveling thousands of miles. (Chapters 2 and 15 discuss how HF radio signals travel.) HF marine radio works particularly well at sea, where the salt water acts as a superb radio mirror, enabling the signals to travel great distances.

Using FM is not efficient on HF bands, so *single-sideband* (SSB) signals are used. SSB is a type of AM signal as discussed in Chapter 5 on Citizens Band. By convention, all marine SSB communications use the higher frequency of the two sidebands, or *upper sideband* (USB). As with the marine VHF radio channels, the HF channels follow a system that covers the wide range of frequencies (see Table 8-2).

The ability of the ionosphere to reflect signals changes through the day, the season, and with the solar cycle, so you need a good selection of frequencies for reliable communications. You can find literally dozens of HF frequencies that are used by mariners, so only the primary HF channels are listed in Table 8-2.

Table 8-2	Primary HF Marine Communication Channels
Band	*Channel Frequencies (kHz)*
4 MHz	4146, 4149, 4417
6 MHz	6224, 6227, 6230, 6516 (daytime)
8 MHz	8294, 8297
12 MHz	12353, 12356, 12359, 12362, 12365
16 MHz	16528, 16531, 16534, 16537, 16540
18 MHz	18825, 18828, 18831, 18834, 18837, 18840, 18843
22 MHz	22159, 22162, 22165, 22168, 22171, 22174, 22177
25 MHz	25100, 25103, 25106, 25109, 25112, 25115, 25118

Knowing your distress channels

Just as with marine VHF radio, marine HF radio has designated channels just for distress and emergency calls. DSC is also used on HF marine radio, so you have access to two sets of HF distress channels, one for voice, and one for DSC. Here are the distress channels:

- ✔ **Voice:** 2182, 4125, 6215, 8291, 12290, 16420 kHz
- ✔ **DSC:** 2187.5, 4207.5, 6312, 8414.5, 12577, 16804.5 kHz

Operating a marine HF SSB radio

Because marine HF channels cover such a wide frequency range and SSB (single sideband) operation requires more operator involvement than FM radio, you can find even more controls with HF SSB gear than with other marine VHF radios.

HF SSB communications are more susceptible to noise and are likely to involve weaker signals than on short-distance VHF FM. This is not unexpected — the signals are likely traveling hundreds or thousands of miles!

You may recognize the radio controls shown in Figure 8-2. However, because those who use HF radios refer to operating frequency directly, the display on these radios shows both the channel number and the operating frequency.

Here are some other characteristics of HF radios:

- ✔ **Memory Group Select:** You can store and group frequencies in the radio's memory according to the needs of the traveler. For example, you might have one memory group dedicated to your home region, another to the destination, and another full of channels you use on the open sea.

- ✔ **Antenna Tune:** This control makes the antenna tuning equipment adjust for optimum power transfer from the transmitter to antenna.

- ✔ **Keypad:** Use the keypad to quickly and directly enter the operating frequency to which you want to switch.

- ✔ **Mode:** Although HF voice communications use upper sideband (USB), the radios can also send and receive data by using a *frequency-shift keying (FSK* or *AFSK)* mode. CW is an abbreviation for the mode used for sending and receiving Morse code, a great backup mode under difficult conditions.

- ✔ **NB (Noise Blanker):** A system that clips out the buzz or rasp of noise generated by engines and motors. The noise blanker compromises receiver selectivity to some degree, so use it only when you need it.

- ✔ **AGC (Automatic gain control):** A circuit that adjusts the receiver sensitivity *(gain)* so that a relatively constant volume is produced. You may need to turn this circuit off when a nearby strong signal is causing a desired weaker signal to be suppressed or covered up.

- ✔ **Transmit Frequency:** Similar to the Monitor buttons on FM radios, this control disables the squelch circuit to listen for a weak signal.

- ✔ **Distress Channel and Call**: Prominently located, these buttons allow the operator to quickly initiate a standard distress call.

Signal quality on marine HF radios can vary quite a bit — much more than it does on marine VHF radios, where signals are either fairly strong or unreadable. On marine HF radios, quality can be very good to nil and everything in between — sometimes within minutes — because of variations in the signal path. To accommodate these changes, you need to practice using your radio's controls.

It takes a fair amount of experience to become a skilled HF marine operator. An excellent reference to help you find out more about the HF SSB service is Gordon West's e-book, *Marine SSB Simplified,* which you can download for free from the Icom Web site: www.icomamerica.com/downloads/#pdf.

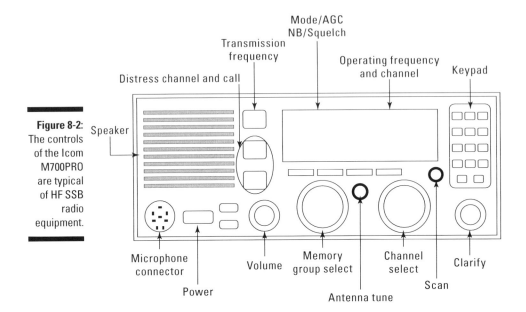

Mode/AGC
NB/Squelch

Transmission
frequency

Operating frequency
and channel

Keypad

Distress channel and call

Figure 8-2:
The controls
of the Icom
M700PRO
are typical
of HF SSB
radio
equipment.

Speaker

Microphone
connector

Power

Volume

Memory
group select

Antenna tune

Channel
select

Scan

Clarify

Selecting a marine HF radio and antenna

Unless you're familiar with boats and radios, work with an expert HF radio installer. That person is qualified to determine which equipment is likely to work on your boat and will perform the communications that you need. HF radio is really a *radio system*. All the pieces need to work together to be effective (see Figure 8-3).

The antenna tuner is required because of the wide range of frequencies used in HF SSB communication. At the lower frequencies, the antenna is very short compared to the signal's wavelength and the opposite at the higher frequencies. Electrically, this causes the antenna to look very different to the transmitter at different frequencies.

The antenna tuner is used to *match* the antenna and transmitter, allowing maximum power output.

If your boat has a mast, an antenna can be strung between the mast and either the bow or stern. The *insulated backstay antenna* in the figure is very popular and does double duty as a structural component. An HF whip antenna (a larger version than those used with marine VHF) can be mounted on the fantail or bridge of a powerboat. In both cases, the ground system is important and the installer will need to customize the installation to the particular characteristics of your boat.

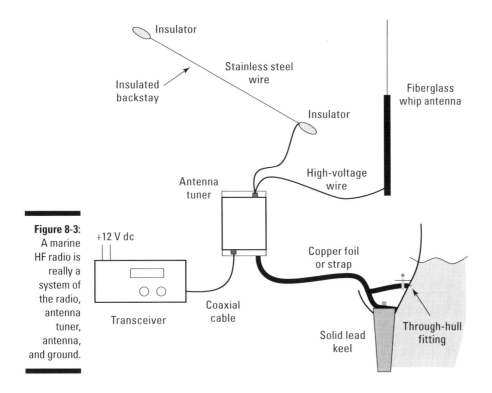

Figure 8-3:
A marine
HF radio is
really a
system of
the radio,
antenna
tuner,
antenna,
and ground.

In a ship, ground means *water*. If the boat has a metal hull, radio systems can use that because it's in direct contact with a lot of water. Wood or fiberglass hulls have *through-hull* fittings for various functions that can be used for water contact. Often, several such fittings can be connected together with copper foil or strap to make a good water connection. If the boat has a solid lead keel, that can also be used.

Basic Marine Radio Do's and Don'ts

Radio etiquette is imperative whether you're at home, in your car, or traveling the open seas. In fact, most of the advice I offer here is applicable to your other radio adventures. Here are the do's and don'ts:

✔ **Do keep your communications simple and direct.** The person on the other end has to listen to you through a small speaker and all the ambient noise.

Most marine radio contacts are very short — reporting status or plans, inquiring about status, or doing some kind of radio errand. For longer chats, use a CB radio or a cellphone so that you don't clog the limited channels. Ham radio is perfect for keeping in touch with friends while out and about, too.

✔ **Do listen.** When you're not using the radio, listen to other mariners make contact and pass information.

✔ **Don't start talking right away.** Before transmitting, listen to the channel for at least ten seconds. If you're not sure whether the channel is in use, ask — you might not be able to hear both sides of a conversation.

✔ **Do know your identity.** Your marine radio identity is your boat's name, so get used to identifying that way.

✔ **Don't sound like a *poseur.*** Say, "Over," to indicate that you're done talking and are going to release the PTT switch. When the conversation is done, identify and say, "Clear." If you say, "Over and out," you will sound inexperienced. Sounds good in the movies, but not on the real airwaves.

✔ **Don't ask, "Do you copy?" or "Do you read me?"** These phrases aren't needed because the request is implicit in your transmission.

Performing Basic Radio Tasks in the Water

You need to know how to do some basic functions with your radio, especially before you take it out on a trip. Practice these basics so that you sound like a pro on the water:

✔ **Checking your radio:** Asking for a *radio check* is a common way to test your radio system. On marine VHF, use Channel 16 or Channel 9. On HF, select any of the frequencies that you think should support communication. Say the name of your vessel when you do your check. For example, if your boat is called the Pleasure Barge, say, "This is Pleasure Barge for a radio check, please." Wait for a reply — you may have to try a few times to get an answer.

You may be asked to perform a *five count* — identify your boat and count over the air from one to five and back again. This gives the responding station time to evaluate your signal. Be sure to say thanks!

✔ **Contacting another vessel:** Say, "Calling [the other vessel's name], this is Pleasure Barge." Give the other boat's name first to attract the operator's attention. If you get no response, wait for ten seconds or so before trying again. After three tries, you can assume the person you're looking for probably isn't listening.

All vessels are required to keep a watch on the calling and distress frequencies. On marine VHF, that's Channel 16. Monitor the channel whenever you're at the radio. Your obligations to keep an emergency watch are discussed fully on the U.S. Coast Guard's Web site at www.navcen.uscg.gov/marcomms/watch.htm. Keeping watch on the HF distress channels (listed earlier in this chapter) is not required, but is a good idea anyway. Monitoring one of the lower frequency channels enables you to hear nearby vessels. Make monitoring the channels a habit — one day you might save a life!

Satellite Radio and Marine E-mail

Just as foreign news correspondents can report breathlessly from the center of the action, no matter where that may be, you have access to the very same technology that can phone home from nearly anywhere on the planet. Satellites orbiting the earth provide connections to handheld satellite phones (often called *sat phones*), laptop-sized terminals, and radio-sized satellite communications *(sat comm)* systems. The best-known service providers are:

- ✔ Intelsat (www.intelsat.com)
- ✔ Globalstar (www.globalstar.com)
- ✔ Orbcomm (www.orbcomm.com)
- ✔ Iridium (www.iridium.com)

Marine communications companies act as distributors for these systems so that you can buy or lease equipment and pay for satellite time just as you would with a mobile phone account. A number of different access plans are available, some of which include data connections so that you can send e-mail or transfer other data content. In most cases, you pay by the minute, and it's a pretty penny — data connections cost more than satellite phones and the charges increase based on the speed of the data transfer you request.

Of course, e-mail and other data services are available without going through a satellite. The best known service for personal use is SailMail (www.sailmail.com), a system of land-based stations around the world that you can use for $250 per vessel per year (this price is likely to change, of course). Vessels use an HF SSB radio and a special data *modem* or data transceiver. Figure 8-4 shows the basic system. A modem translates digital data into audio frequencies that can be transmitted over a voice channel. Compared to regular Internet speeds, SailMail is relatively slow, good for a quick message check, to order parts, or to make arrangements for accommodation at the next port.

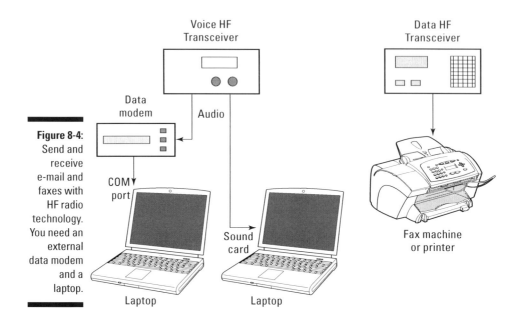

Figure 8-4:
Send and receive e-mail and faxes with HF radio technology. You need an external data modem and a laptop.

To use SailMail, the traveling vessel tunes a radio to one of the listed frequencies and initiates a process by which the boat's data modem attempts to connect to the shore modem. If the channel isn't busy and the signal is of adequate strength, the two modems allow a host computer running a program called AirMail to exchange messages with a SailMail e-mail server. The system uses lots of frequencies and stations, so you can connect from almost any location. You don't have to be afloat — SailMail works quite well over land, too. You can get more detailed information at the SailMail Web site.

Ham radio licensees (see Chapter 9) use a similar system called WinLink (winlink.org) that also uses the AirMail software. Hams have even wider coverage than SailMail, but you can't send business e-mails via ham radio.

Another HF data service available for the cost of a radio is *weather fax*. Shore stations around the world transmit a continuous sequence of weather charts as facsimile transmissions. If you can hook up a fax machine or laptop running fax software to your radio, the information is yours at no charge. The details are available on the HF-FAX Web site, www.hffax.de.

Getting That License

If you are operating a personal pleasure craft near the U.S. coastline or in domestic waters, you can operate on marine VHF, install an emergency beacon, and even run marine radar without a license. (Regulations vary some from country to country.) To operate in international waters or to use HF SSB or data, you need to get the proper license. Licenses are the same for any size or type of vessel. You'll find complete information on marine radio licenses at `wireless.fcc.gov/marine/fctsht14.html`.

Actually, you need not one, but two licenses, each of which is good for ten years; you need one for the vessel itself ($205), called a *Ship Station License,* and one for each operator ($55), called a *Restricted Radiotelephone Operator Permit.* No exam is required for either license. The process is detailed at `www.marinecomputer.com/articles/licensing/licensing.html`.

The licenses are good around the world by international treaty. When you go through the licensing process, you obtain your vessel's MMSI number (see the section on DSC earlier in this chapter). Don't think you can bluff your way along without a license. You'll need a valid *call sign* to get access to many services.

Chapter 9

Citizen Wireless: Amateur Radio

In This Chapter

▶ Getting familiar with amateur radio

▶ Listening to ham radio in action

▶ Obtaining an amateur license

*A*s you gain expertise with your radio, be it a low-power FRS handheld or a more powerful business radio, you may occasionally make an unexpected contact or find yourself taking advantage of a high spot to communicate over long distances. I'll bet you'll think, "Hey, this is kind of fun!" You're right, it *is* fun. But the restrictions placed on most services force you to keep contact focused on the business at hand. This is where amateur radio comes in. Amateur radio provides a creative outlet for individuals just like you to experiment with. You can use radio in incredibly varied ways; if you don't believe me, ask the more than 3 million hams around the world. I've been a ham for more than 30 years and it's given me a lifetime of interesting and challenging activity!

In this chapter, I give you a view from 30,000 feet of the world of amateur or *ham* radio. If you'd like to know more, I wrote *Ham Radio For Dummies,* a desktop reference for the potential and newly licensed ham.

Careful — ham radio is addictive! If you're looking for online information, check out the Web site of the world's largest and oldest amateur radio organization, the American Radio Relay League (the ARRL) at www.arrl.org. A great portal site with dozens and dozens of links to all kinds of ham-related goodies is www.ac6v.com. This site is a good example of ham radio's hams helping hams ethic.

Tuning In Ham Radio Today

Amateur radio, otherwise known as *ham* radio, can be technical, so it's always a good idea to look at the operating procedures and technology, whether you plan to use ham radio for a specific purpose or just to have fun.

Ham is an old term whose origins are obscure. The most likely meaning was a derogatory reference to poor operators from professional telegraphers. Amateurs adopted it as a badge of honor and keep the name today.

A distinguishing feature of amateur radio is that hams can design and build their own equipment or assemble a station from factory-built components without limits on the arrangement of the equipment. Hams delight in the DIY (do-it-yourself) ethic known as *homebrewing* and help each other out to build and maintain their stations.

In other services, the use of the radio is restricted to certain purposes, but not ham radio. You may talk to whoever you want anywhere on earth, in the sky, or even in space. You may converse, using whichever of the many authorized types of signal you choose, from venerable Morse code to the latest digital scheme. Your power level can range up to the legal limit — 1.5 kilowatts in most cases. You can put up antennas wherever you think they'll work and operate in any manner that you enjoy, as long as you respect the rights of other hams.

Ham radio core values

At the core, however, the Amateur Service exists for five reasons, according to the FCC:

- ✔ To recognize ham radio's exceptional capability to provide emergency communications.

- ✔ To promote the amateur's proven ability to advance the state of the radio art.

- ✔ To encourage amateurs to improve their technical and communications skills.

- ✔ To expand the number of trained operators, technicians, and electronics experts.

- ✔ To promote the amateur's unique ability to enhance international goodwill.

Pretty lofty ideals, but that's what hams do, even when they're just on the air having a good time. Want to know more about the most powerful radio service available to private citizens? Read on!

Common ham radio activities

If you tune a radio across the ham bands, here are some of the things you might hear hams doing (this list is by no means comprehensive):

- ✔ **Talking:** By far the most common type of activity for hams, engaging in conversation is no small deal. You can talk to people across town or across continents and oceans. You don't have to know another ham to have a great chat — ham radio is a very friendly hobby with little class snobbery or distinctions. The thrill is in making contact with strangers who share your interest. So tune in and start talking!

- ✔ **Networking, sort of:** Hams frequently engage in scheduled contacts and organized groups called *nets* (short for networks). Many hams keep repeating schedules to keep in touch with family and friends. Nets are organized on-the-air meetings for hams who share an interest or purpose, such as astronomy, caving, railroads — you name it and there's a net for it. *Traffic nets* are organized to exchange messages and connect with other such nets in different areas to pass messages around the world. Other nets are more like bulletin boards for swapping equipment, discussing special interests, or even playing a game of chess by radio. You find a master directory of nets in the American Radio Relay League's (ARRL) Net Directory, www.arrl.org/FandES/field/nets.

- ✔ **Emcomm:** Emergency communication has been part of amateur radio since its earliest days. In fact, it's one of the fundamental reasons that governments make room for an Amateur Radio Service (ARS) in their allocations of the radio spectrum. Following natural disasters or in any time of need, organized groups and individual amateurs fill in until the usual lines of communication (telephone, television, and so on) are restored or replaced. You can find many emergency communications teams around the United States organized at the county level by the ARRL's Amateur Radio Emergency Service (ARES). You may hear training and drill activities on weekends as hams practice their emergency skills.

- ✔ **Contests:** On some weekends, a few of the ham bands may fill with rapid-fire contacts. A ham radio contest is underway! Contests are events in which hams try to make as many contacts as possible — sometimes in the thousands over a weekend — exchanging short messages with other hams. Along with contests, there are thousands of special event stations and awards that are available for various operating accomplishments, such as contacting users in different countries or states.

Using electronics and technology

Ham radio is full of electronics and technology. To start with, transmitting and receiving radio signals can be a very electronics-intensive endeavor, if you wish. When you open the hood of your radio, you're exposed to everything from basic direct-current electronics to cutting-edge radio-frequency techniques including the very latest in digital signal processing and computing.

Voice and Morse code communications are still the most popular technologies by which hams talk to each other, but computer-based digital operation is quickly gaining loyal fans. The most common home station configuration today is a hybrid of the PC and radio. Some of the newer radios are exploring *software-defined radio (SDR)* technology that reconfigures the radio's signal generation and decoding software on the fly, depending on what signals are being exchanged.

Hams also develop their own software and use the Internet with radios to create novel hybrid systems. For example, by combining Global Positioning System (GPS) radiolocation technology with the Web and mobile radios, the Automatic Position Reporting System (APRS) was invented by hams. APRS, by which vehicles report their position so that it can be viewed on a map or tracked by a computer, is now put to work by many different types of users.

Taking full advantage of the Internet, hams have created the Winlink2000 e-mail system, which lets hams use their radios to connect to gateway stations that bridge the radio-Internet divide. Winlink2000 is used for both recreation and emergency communications.

Along with the equipment and computers, hams become students of antennas and *propagation,* the means by which radio signals bounce around and get from place to place. Hams investigate solar cycles, sunspots, and how these astronomical features affect the earth's ionosphere. Weather takes on a whole new importance, generating fronts and disturbances along which radio signals can sometimes travel long distances. Antennas, with which signals are launched to take advantage of all this propagation, provide a fertile universe for the station builder and experimenter. Antenna systems range from small patches of printed circuit board material to multiple towers festooned with large rotating beam antennas.

Experimentation is a hotbed of activity for hams. New designs are created every day and hams have contributed many advances and refinements to the antenna designer's art. After you're licensed, all you need to get started is some imagination, wire, a feedline, and a ham radio.

Whatever part of electronic and computing technology you most enjoy, you can find it used in ham radio somewhere . . . and sometimes all at once!

Finding the Ham Bands

If you have used an FRS/GMRS, MURS, marine, or business radio, the ham bands are right next door! Ham radios can be thought of as the more flexible cousins to the commercial and unlicensed radios. In fact, many hams convert these radios to use on the amateur bands.

Can they do that? You bet! Hams have freedoms to tinker that no other service can claim. Here's a look at the various bands and a tip on figuring out just where those wily hams hang out.

Finding shortwave hams

The frequencies in Table 9-1 are the traditional amateur bands that have been used by hams for nearly 100 years. If you have a shortwave receiver that can receive *single sideband* or *SSB* signals, you can tune in quite a few amateur signals, maybe even from other continents. Ham bands are referred to by wavelength, just like shortwave broadcast bands. (LSB and USB refer to lower and upper sideband modes, which I discuss in Chapter 15.)

Table 9-1	The U.S. Amateur Shortwave Bands	
Band	*Morse and Digital Signals*	*Voice Signals*
160 Meters	1.800 to 1.860 MHz	1.843 to 2.000 MHz (LSB)
80 Meters	3.500 to 3.750 MHz	3.750 to 4.000 MHz (LSB)
60 Meters	Not permitted	5330.5, 5346.5, 5366.5, 5371.5, 5403.5 kHz (USB)
40 Meters	7.000 to 7.150 MHz	7.150 to 7.300 MHz (LSB)
30 Meters	10.100 to 10.150 MHz	Not permitted
20 Meters	14.000 to 14.150 MHz	14.150 to 14.350 MHz (USB)
17 Meters	18.068 to 18.100 MHz	18.110 to 18.168 MHz (USB)
15 Meters	21.000 to 21.200 MHz	21.200 to 21.450 MHz (USB)
12 Meters	24.890 to 24.930 MHz	24.930 to 24.990 MHz (USB)
10 Meters	28.000 to 28.300 MHz	28.300 to 28.600 (USB)

Worldwide, the ham band frequencies are very similar to these, except that voice signals outside the U.S. are common in the Morse and digital frequency ranges.

VHF, UHF, and microwave signals

Above 30 MHz, the ham bands are found next to or near the business bands. Like nearly all VHF and UHF radio users except aviation, ham communications on these bands is mostly FM on regularly spaced channels. Scanners like those in Chapter 10 can be easily programmed to scan these channels, along with police and fire department channels. Table 9-2 lists the ranges of frequencies in which a scanner in search mode will find amateur activity. Hams also use repeaters, so you can find a mix of strong repeater outputs and weaker input signals from individual hams.

Table 9-2	Amateur FM Operation on VHF and UHF
Band	*Frequencies*
6 meters	51.12 to 53.98 MHz
2 meters	145.2 to 145.5 MHz
	146.00 to 147.99 MHz
1¼ meters	223.85 to 224.98 MHz
70 cm	442 to 450 MHz
30 cm	1282 to 1288 MHz

Hams are also active on microwave bands at 2.4 GHz and higher. These bands require a special receiver, but if using microwaves piques your interest, the hams are there, using signals ranging from Morse code to wireless networks and lasers!

Getting a Ticket: The Ham Kind

How do you get an amateur radio license? Is it hard? What about Morse code? How much does a license cost? Are there different kinds of licenses? You have plenty of questions, I am sure. A ham radio license does require you to take a test, but despite this fact, getting a license is very straightforward. Although getting a ham license requires more work than just filling out a form and sending in your money, the process is designed to help you get your license and get on the air.

Understanding why an exam is required

You may have thought your test-taking days were behind you. But if you want to use a ham radio, you need to take at least one more exam. Why? Well, there are a couple of reasons:

- ✔ **Regulation:** Amateurs communicate internationally and directly, without using any kind of intermediate system that would regulate their activities.

- ✔ **To avoid chaos:** Because of the broad scope of amateur radio, hams should have a minimum amount of technical and regulatory background so they can coexist with other radio services, such as broadcasting.

This combination of technology and independence is pretty powerful so, by international treaty, hams are required to demonstrate that they understand the rules and the basics of the technology.

Preparing for the exam

The ARRL and other organizations publish study guides and manuals to help you study for the test. Some books may be available through your local library. (Be sure they cover the latest version of the exam, because the test questions change from time to time.) Online tests are available with actual test questions.

Exams are administered by amateurs certified as *VEs* or *Volunteer Examiners*. You don't have to go to a government office or anything like that to be a VE. I'm a VE and I give the exams in my kitchen! You can also take the test through a ham radio club or at a ham radio convention or swap meet. The testing is handled in a very comfortable, personal manner and the cost for the entire session, including your licensing fee, is only $14 — less than a large pizza!

Knowing which exam to take

There are three types of licenses; Technician (the usual beginner's license), General, and Extra. You take progressively more challenging exams, starting with the Technician exam, and then moving on to the General and Extra exams, moving at your own pace; as you move through the licensing process, more and more frequencies become available to you at each level.

The General and Extra Class licenses require you to show some slow speed Morse code ability.

After you pass a specific test level, called an *elemenw* you have permanent credit for it. Your license is good for ten years and can be renewed for a nominal fee (currently it costs $14, big spender) without taking an exam.

If this introduction to ham radio has gotten you thinking, find a local club on the ARRL Web site and start asking questions. I hope to put you in my logbook someday!

Part III
Listening In: Scanning and Shortwave Listening

The 5th Wave   By Rich Tennant

In This Part . . .

With all the activity on the airwaves, you shouldn't be surprised that a community of listeners has sprung up around the world. The biggest group by far is the scanners, who keep an ear on what's happening in their towns and regions. The capabilities of modern scanners are light years beyond what was available just a few years ago, so you are in for some good reading in Part III.

To really enjoy your scanner, you need to understand how to listen to the various types of radio systems. I start by covering the most-listened to systems on the airwaves — public safety and service agencies. You can track and monitor sophisticated computer-controlled systems with your scanner, but only if you know how.

Listening to the comings and goings of the aviation community is also fascinating. Part III shows you how to sit with the pilots and controllers as they go about their daily flights. Military and government personnel are out there on the airwaves, too, so I introduce you to their radio systems and frequencies. Finally, a scanner gives you a new window on public events, such as auto racing, fly-ins, parades, and more. Learn how to make a scanner part of these recreational activities.

Along with local and regional goings-on, shortwave listeners tune in to stations from around the world. There's so much more to broadcasting than the local AM and FM stations. You can also find out about the shortwave channels for government and commercial operation, including data and text services. With receivers packing amazing performance in handy, portable packages, shortwave listening has never been easier. Tune in and find out how you can listen to the world turning!

Chapter 10

One Adam 12: Scanner Basics

This chapter is all about scanners, those special radios that enable you to listen to all the chatter in the radio world around you. Today's scanners can soak up the thousands of different conversations and events happening in cities and towns, big and small. It's all there, literally at your fingertips.

Why scan? Aside from curiosity, listening to your local public safety agencies can make you feel more secure and less uncertain. Over time, you'll also gain a new appreciation and understanding of the work these departments do every day. As you gain familiarity with your scanner and signals, your curiosity will lead you to broaden your horizons to other services and channels. You can find an unlimited number of signals out there.

This chapter discusses the common signals and services so that when you start operating your scanner you know more about why the controls work the way they do. I also touch on the programming features of today's scanners. Finally, I discuss choosing scanners and antennas. In subsequent chapters, you can find out more about the major types of services and how to use your scanner to listen to their activities.

Listening: Oh, the Signals You'll Hear

You have to address a few basic radio notions before you can fully enjoy the scanner experience. By understanding the basics, the information in your user's manual will make much more sense and you'll be plucking signals from the sky in no time. If you're already a veteran scanner, you can skip ahead a couple of sections.

Getting scanner basics

If you look closely at your radio, you'll find a button labeled SCAN on many different types of radios. Does that button make the radio a scanner? Not really. Those radios offer a limited version of a scanner's true functionality. A *scanner* is a special type of receiver that is designed expressly to rapidly tune across a series of frequencies and alert the user when communications are found. Modern scanners can process a large number of different frequencies at blazing speed.

Introducing channels and services

The radio spectrum is spread out over a huge range of frequencies and wavelengths (see Chapter 2 for more information on this topic). It may seem impossible to pick just one needle out of that huge metaphorical haystack, and if you had to turn a dial to find a desired signal it probably would be! Because many smart people were on the job, signals aren't just randomly sprinkled about; they're lined up in an orderly sequence of frequencies called *channels*. A channel is associated with a specific frequency and *channel spacing,* which is the difference in frequency between adjacent channels. A radio signal occupies a small range of frequencies and must stay within a *channel width* to avoid causing interference to its neighbors.

Similar types of signals and users are grouped together as *services.* (See Chapter 2 for more information about radio services.) By knowing what channels are allocated to what services, the task of finding a desired signal is greatly simplified.

How radio signals get from place to place depends on their frequencies. For example, lower-frequency signals tend to travel a lot farther than their higher-frequency cousins. Those same smart people decided to put short-range users on short-range frequencies and vice versa. Clever, eh? The ranges of services are allocated in *bands* of frequencies, as shown in Table 10-1. Not all scanners receive all bands. This layered organization of the radio spectrum — band-service-channel — streamlines finding the signals you want to listen to.

Table 10-1	The Basic Radio Bands		
Frequency Range (MHz)	*Band*	*Frequency Range (MHz)*	*Band*
25 to 26.960	Businesses and Shortwave Broadcast Band	174.0 to 216.0	TV High VHF Broadcast Band
26.965 to 27.405	Citizens Band	216.0 to 222.0	Business Band
27.410 to 27.995	Mobile and Fixed-Location Stations, called Business Band	222.0 to 225.0	1.25-Meter Amateur Band
28.0 to 29.7	10-Meter Amateur Band	225.0 to 328.6	Business Band
29.7 to 49.990	Low VHF Band	328.6 to 335.4	Aviation Air Band
50.0 to 54.0	6-Meter Amateur Band	335.4 to 405.9875	Business Band
54.0 to 72.0	TV Broadcast Band	406.0 to 419.9875	Federal Government
72.0 to 76.0	Business Band	420.0 to 450.0	70-cm Amateur Band
76.0 to 88.0	TV Low VHF Broadcast Band	450.0 to 469.9875	UHF Business Band
88.0 to 108.0	FM Broadcast Band	470.0 to 606.0	UHF TV Broadcast Band
108.0 to 117.975	Aircraft Navigation	606.0 to 614.0	Business Band
118.0 to 136.975	Aviation Air Band	614.0 to 698	UHF TV Broadcast Band
137.0 to 143.9875	Military Band	698.0 to- 806.0	Business Band
144.0 to 148.0	2-Meter Amateur Band	806.0 to 960.0	Public Service and Mobile Phone Band
148.0 to 161.995	VHF High Band	960.0 to 1240.0	Aviation Radionavigation and Radiolocation
162.0 to 174.0	Federal Government Band	1240.0 to 1300.0	23-cm Amateur Band

Joining a scanner club

There are literally millions of folks out there using scanners. Why not take advantage of their expertise by joining an online scanner club? The RadioReference Web site (www.radioreference.com) is one of the most widely used sites on the Web for exchanging information about scanners, scanning, and radio channels. You can find many other clubs by conducting a Web search for the term *scanner club*. The Yahoo! Groups Web site also has many scanning-related groups. From the Groups home page (groups.yahoo.com), click Hobbies & Crafts. Then click Hobbies. From there, look for the Amateur and Ham Radio link.

I urge you to join one or two of these groups as you seek info about scanners. There is only so much information I can provide in this book! After you become a scanning veteran, you'll be able to help other newcomers over their hurdles, as well!

This table doesn't list all the individual services in each band, but it gives you an idea of what can be found in different frequency ranges in the United States. Internationally, you may notice some variations in the assignments, particularly above 50 MHz. Many of these bands are shared between the different services and users.

A scanner takes advantage of this radio regimentation by being able to jump from channel to channel much faster than a *communications receiver,* which is designed to tune to any frequency. (Communications receivers may have scanning functions, of course.) Because radio users tend to use groups of closely related frequencies within one service, scanners can further organize the channels into *systems*, such as a local public safety system.

Using AM, FM, and digital signals

As you might imagine, several types of signals are wafting around the airwaves. On the bands you'll be listening to, the vast majority are voice signals transmitted by FM or *frequency modulation*. (Chapter 2 talks about the difference between the various types of modulation.) Most systems use *wideband* FM signals that occupy up to 25 kHz of spectrum space. To utilize the radio spectrum more efficiently, some systems have changed to *narrowband FM* or *NFM* signals that only take up 12.5 kHz. You have to set your scanner to the correct mode to receive NFM signals properly; otherwise, they'll sound weak.

AM signals (including *single sideband* or *SSB* and *Morse code*) are not used as often as FM signals, but you can find them on the aviation frequencies and on portions of the amateur bands.

Digital signals convert the speaker's voice to a stream of codes that sound like a rasp on an AM or FM receiver when they're transmitted. Digital voice transmissions are not yet widely used, but I expect digital voice to become much more common over the next five years. More and more scanners include digital capability, so if you plan on using your scanner for a while, getting one with digital capabilities is important.

Simplex communications and repeaters

The simplest way of communicating is to use just one frequency, taking turns transmitting and receiving on it. That's *simplex* communications and it's very common for short-range radios, such as FRS/GMRS radios and Citizens Band. The range of simplex communications is a few miles at best. When you listen to simplex communications, you may not hear both stations equally well because they are in different locations.

It would be too difficult to equip all mobile stations with a powerful enough radio to enable all users to hear each other. The solution is to use a *repeater,* which listens on one frequency and retransmits what it hears on another. (I describe repeater systems in Chapter 4.) Scanners usually just listen to the repeater's *output* frequency, but it's handy to be able to listen to the repeater *input,* too. When you listen to repeater communications, the signal from the repeater is the same strength no matter who's transmitting, because you're just listening to the output of the repeater's transmitter. The individual stations are actually transmitting on a different frequency where the repeater receiver is listening.

Dispatch versus one-to-one communication

Many businesses and government agencies use radio systems to communicate between one or two stations and many mobile users. A good example is a local trucking company. Orders for deliveries are constantly streaming into the home office and being relayed to drivers out on the streets. This is a *dispatch* or *fleet dispatch* system. The station at the home office is called a *base* station and its operators are called *dispatchers.* The moving stations are called *mobiles* or *subscribers.* By far, this is the most common type of radio usage today. When you listen to these systems, you can hear the dispatch station talking frequently, giving orders, taking reports, and so on. The mobile stations transmit in response to the dispatcher or when they need information.

Communications that don't require a controlling dispatcher use a *one-to-one* communication system. Examples include firefighters talking to each other directly at a fire, pilots coordinating directly with each other while in the air, or amateur operators sharing a repeater. Figure 10-1 illustrates the different types of radio use. One-to-one systems may be simplex or repeater communications.

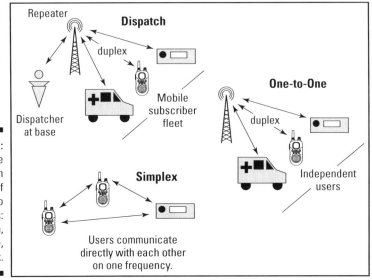

Figure 10-1:
The three common types of radio systems: dispatch, one-to-one, and simplex.

Introducing the Radio Population

As you scan through the active channels in your area, you'll meet several different types of radio users. Each has different procedures and communicates different types of information. Depending on your interests, you may want to narrow your focus to one type of user. If you're new to scanning, allow me to introduce some of the folks you'll be getting to know.

Business users

Businesses make up the largest number of radio users. Large businesses, such as trucking firms or taxicab companies, may go ahead and license a radio system for their exclusive use. On the other hand, a smaller company may use a *shared* system maintained by a communications company for its *subscribers,* sharing the costs and benefits of the system among them.

With all the new technology out there, you might wonder why businesses don't use mobile phones to communicate, instead. In fact, many of the smallest businesses and companies that require direct contact between customers and employees often do use mobile phones rather than radios. Large businesses, however, find it difficult to manage such a fluid system. It is also more difficult to communicate *one-to-many*. When a message needs to go out to all mobile stations immediately, mobile phones just won't do the trick.

Government users

City, state, and federal government users make up another substantial portion of the radio population. This group includes public safety as well as utility, academic, and transportation agencies. Government users employ systems that are very similar to those used by businesses, but are kept apart from business users in separate bands to prevent private companies and public agencies from being in direct competition for radio spectrum.

Military users

Military users, communicating at and around military bases are another part of the radio population. Military radio systems use the same radio signals as those used by government and business users. As a result, you can receive the signals with your scanner. Sensitive military signals are *encrypted* or *scrambled* so that they can't be intercepted by the casual (or hostile!) listener. However, most messages are transmitted as *clear speech*.

Hobbyists and other individual users

There are many opportunities for individuals to make use of radio. Boaters make heavy use of the Marine VHF channels in the VHF High Band. Refer to Table 10-1, to see the Citizens Band allocation, as well as numerous amateur radio bands throughout the spectrum. The popular FRS/GMRS radios are available for individual use in the UHF high band. You can use your scanner to listen to these busy frequencies, too.

Public and private aviation users

Aviators rely heavily on radio to coordinate activities in the air and on the ground. Radio is so important to aviation that it has its own dedicated bands. Aeronautical mobile frequencies are used by pilots for air-to-air and air-to-ground communications. If you live near an airport, you'll find the aviation band to be a source of endless activity. *Radionavigation* and *radiolocation* bands are not used by humans, but by *avionics* (the electronic devices in the airplane that maintain speed, altitude, and so on), which guide the pilots.

Listening to the weather channels

Similar to broadcast stations are the weather channels that provide a continuous stream of weather information for a particular area. Most scanners can receive these channels and many even have dedicated controls to access them quickly or tune to them more frequently than other channels. In areas of the country with frequent severe weather, the weather channels are very important. The National Oceanic and Atmospheric Administration (NOAA) Web site (www.nws.noaa.gov/nwr/index.html) gives information about the weather channels and the new SAME (Specific Area Message Encoding) technology. The National Weather Service (NWS) uses SAME technology to provide targeted information for your area. You can program your scanner to receive information and alerts. Pretty nifty stuff.

Learning How to Use A Scanner

The best way to find out how to operate your scanner is to monkey around with it. By following a few simple steps, you can get started right away, finding frequencies, picking up signals, and listening in. Most scanners come with a set of channels already programmed in, so put in some batteries, turn it on, and you should be ready to roll! The following set of steps describes what the scanner's receiver is doing when it's jumping from channel to channel:

1. **Tune the scanner to a channel frequency.**

   The channel may be locked out, so the scanner skips it and goes on to the next channel.

2. **If the channel is locked out, tune to the next frequency.**

   The scanner keeps tuning until it gets to a channel that's not locked out.

   Is there a signal present?

3. **If so, turn on the speaker and wait for the duration of the delay period.**

4. **If no signal is present, tune to the next channel.**

   The scanner just keeps going until it finds a channel with a signal. Let the good times roll.

Obviously, there is a lot of technology behind the front panel that makes all this good stuff happen. But all you have to do is tell the scanner what channels to scan and what to do when it finds a signal. The following sections give you some more details about the various controls and functions available with most scanners.

Handling basic controls and use

Even with all the radio technology crammed inside, a scanner doesn't need very many controls. Most of the time, all you'll do is turn it on and turn it loose.

Figure 10-2 shows the layout of the controls for the popular portable scanner from Uniden, the BC246T. Controls for other desktop scanners are very similar. You may notice that Function (FUNC) and Menu buttons are located on the side of the scanner.

Figure 10-2:
The Uniden BC246T scanner has a simple set of controls. Access more complex features with the keypad and menu options.

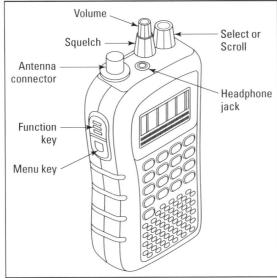

Taking control of the power button

Some scanners combine power with a volume control; others have a dedicated power switch or button. The standard symbol for power is a circle with a line that goes partway through the top (it looks a little like a clock face set to 12:00) and power buttons are generally orange. The power button may be on the front panel or on the side. You may need to press the power key and hold it for a couple of seconds before the scanner turns on or off.

The keys on scanners of all sorts commonly have multiple functions — even the power key! Read your user's manual carefully to see if a function requires press-and-hold or short presses or if multiple short presses mean different things. You can get mighty confused if you accidentally activate an unexpected function.

Adjusting the volume or audio (AF) Gain

On some scanners, the volume is adjusted with keys just like the ones on your TV remote control. Press one button to increase the volume; press another one to decrease the volume.

Modifying squelch settings

The squelch circuit mutes the receiver output in the absence of a received signal, *squelching* the noise. (See Chapter 4 for more about squelch.) Most scanners enable you to adjust the threshold at which the squelch *opens* (turns on the speaker). Raising the threshold so that a stronger signal is required to activate the speaker is called *tightening the squelch*. Lowering the threshold is called *loosening* or *opening the squelch*. Exceeding the threshold to become audible is called *breaking squelch*.

Squelch is an important control because the scanner decides whether a signal is present or not based on whether the squelch threshold is exceeded. If your squelch is set too low or loose, the scanner will stop scanning when no signal is present, only noise. Set too tight or high, the scanner won't detect a valid signal on an active channel.

Adjusting frequency and entering characters

Frequency entry uses numeric keys, although you may have to use a *function* key to start the entry. Some scanners require you to enter a value for every digit on the display; for example, you have to enter 154.0 MHz as 154.0000 before the scanner will accept it. Others fill in the blanks for you based on the band and preset channel step size.

Character entry to create a text label for the channel is more difficult because the scanners don't have keys for each character. For example, CITY FIRE is easier to read and remember than 154.2500. Your scanner may have a char-acter-select mode with a rotary control or up/down buttons to find a letter; then there's another control to select it, and so on. You have to repeat the find-and-select process for each character.

If this manner of character entry doesn't sound like a lot of fun, you're right — and that's why if you have PC you may want to consider using it to set up your scanner.

Starting, suspending, and stopping scans

Look for the Scan key or button. This is usually a touch-and-go control to start and stop scanning. Many scanners also include a Hold or Delay key. This key pauses the scan at the current channel. Using the Scan key to stop and then restart often starts the scan from the first channel, not from the current channel. That may or may not be a big deal, but you should delve into your user manual to see how that works exactly for your scanner.

Locking out channels

If you have dozens of channels and systems (groups of channels) stored away in your scanner's memory, you probably don't want to scan *every* channel *every* time you turn on the radio. To avoid scanning the unwanted channels without erasing these channels from the scanner for good, use the *lock-out* function to skip that channel. This may be a separate key with a label such as Lock, LO, or L/O.

Finding CTCSS and DCS signals

CTCSS stands for continuous tone-coded squelch system, although it's also sometimes referred to as continuous tone-coded sub-audible squelch (rolls off the tongue nicely, doesn't it?). DCS stands for digitally coded squelch. These acronyms refer to additional signals sent along with the user's voice in order to identify them to a receiving station. Your scanner can use them to filter out unwanted conversations. I discuss CTCSS and DCS in detail in Chapter 4.

Understanding how to use your scanner's secondary functions

In order to get the maximum use out of every key and maximize the number of functions available on the keypad, your scanner manufacturer may have assigned more than one function to a single key. These other functions (called *secondary functions*) are usually accessed by using an activating key that acts similarly to the CTRL or ALT keys on a PC keyboard. The scanner key is often labeled FUNC or FNC, as shown in Figure 10-3. You may have to press and hold the key to use the secondary function. Some scanners only require you to press the Function key momentarily; an indicator on the display shows that all keys have their secondary meanings. This is explained in your user's manual.

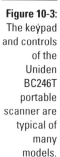

Figure 10-3: The keypad and controls of the Uniden BC246T portable scanner are typical of many models.

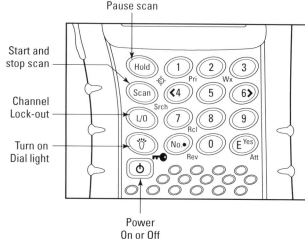

Pause scan

Start and stop scan

Channel Lock-out

Turn on Dial light

Power On or Off

Using your scanner's Priority Scan feature

Of course, a few channels may have more importance to you than others; the local fire or police dispatch, the CB emergency channel, or your business frequency, for example, may be of particular interest. (Living on an island, I listen to the ferry operations channel!) You can assign these channels a *priority* status, to be checked more frequently than other channels. That way, you'll find out immediately if something important or interesting is occurring. You can turn on and off priority scanning any time. In fact, some scanners can enter a priority channels-only mode.

Scanning versus searching

Some scanners have the ability to tune continuously or *search* through a range of frequencies in order to find signals away from known channels. This is a particularly useful feature in portions of the Amateur bands, where because operations aren't *channelized*, activities occur on varying frequencies from day to day. By using the search function, you can also sweep through a range of channels looking for any that may be active. The exact name for this function varies depending on your manufacturer. No matter what it's called, you can find it described in the user's manual for the scanner.

Configuring your scanner

If your scanner isn't all that complex, you may have very few controls other than those on the keypad or front panel. On the other hand, you may have several menus with many options if your scanner's a little more complicated. In either case, you should know about some general settings (called *configuration settings*) that apply to all operations. Here are two of the more common ones:

- ✔ **Scan Delay:** When a scanner finds an active channel, it stops scanning until the transmission ends. The amount of time the scanner waits after the transmission ends until scanning resumes is called *scan delay*. Longer delays allow you to hear more of a conversation before resuming and moving to the next active channel. The default value is typically about one second.

- ✔ **Frequency Step:** Scanners use a default *step size* (how much the scanner changes frequency to get from channel to channel) that may vary with the band in which the scanner is receiving. For example, on VHF High band, the default will be 25 kHz. In some cases, you may want to change to a smaller or larger step. If your scanner has a search function, it may also have a search *step* size that you can specify for searching.

Knowing the Rules of Scanning

You need to know a few important rules for scanning. For one thing, you need to respect the privacy of others. You may be unaware that it is illegal to listen to some types of signals without the consent of the parties involved. These include:

- **Telephone conversations of any sort:** Whether the conversation occurs over a cellular or mobile phone system or you pick up a conversation from a cordless handset of a land line phone, listening to a phone conversation is forbidden.

- **Pager transmissions:** Both analog (voice) and digital pager transmissions are off limits.

- **Any scrambled or encrypted transmissions:** Scrambled and encrypted transmissions are often military or government transmissions. You should move along to the next channel if you run across this kind of transmission.

Furthermore, in some areas, use of portable or mobile scanners outside a home or place of business is either illegal or requires a permit. Why? Because crooks use these devices to listen to the police, that's why! Please check the regulations where you live or plan to travel with your scanner. Call the police department in your area for answers to any particular questions you have. A fairly complete list of all such regulations in the U.S. is available at `www.afn.org/~afn09444/scanlaws`. If you keep your scanner in the car, know the rules for your state and anywhere else you plan on driving.

Trunking Systems

One of the biggest changes in scanning in the past ten years has been the introduction of *trunked* or *trunking systems,* also called *trunked radio systems* or *TRS.* In the old days, all you had to know to listen to the operations of your local police or fire department was a frequency or two. In many regions, those days are gone. The simplicity of single-frequency one-stop shopping is a thing of the past.

Defining trunking

Trunked systems were developed in the 1980s, when the burgeoning radio population began to overwhelm the available radio spectrum. It also became clear that placing different public agencies on different frequencies created problems when the multiple agencies had to work together (such as when responding to disasters). Trunked systems use a computerized central controller to direct individual radios to use one of several possible channels. Trunked systems allow many users and even agencies to share a common set of radio channels, instead of each agency requiring its own channel. The word *trunk* means a common channel shared by different users. Here's how the system works:

Trunked systems have several channels, one of which carries command information to and from the individual radios. This is the *control channel*.

The channels that carry conversations are *data channels*.

Trunked systems organize the users from individual departments, companies, or agencies into groups, each having its own identification or ID.

The controller uses the IDs to direct individual radios to use specific channels. (There are several variations of ID organization, as I discuss in Chapter 11.)

Using your scanner to monitor calls on a trunked system

For your scanner to monitor a trunked system, it must be able to understand the control information used by the system. Each system (such as the Motorola II) has its own specifications. You must also know the IDs of the user groups you want to monitor. With both types of information (system type and group ID) known, the scanner can then follow the radios as they tune from channel to channel under the direction of the controller, as shown in Figure 10-4.

The most popular trunked systems are manufactured by Motorola. Another type is known as *Enhanced Digital Access Communications System (EDACS)*. A variation of EDACS known as *Logic Trunked Radio (LTR)* combines the control and conversation information on the same channel.

The emergence of *digital trunked systems* demonstrates another step in the evolution of shared communications systems. In digital systems, all communications, including voice, are converted to a stream of digital data. You must have a digital scanner to hear these digital transmissions — to an analog scanner (without digital capabilities) the data sounds like the raspy noise of a trunked system's control channel. In North America, the digital standard is *P-25* (short for *APCO-25*) and in Europe, it's called *TETRA*. Check the frequency lists for digital systems in your area before purchasing a digital scanner or the necessary parts to upgrade your current scanner.

Engine 54, your table is ready!

A trunking system is similar to a busy restaurant. The radio channels are the tables; sometimes they're empty and sometimes they're full. The controller is the maitre d', busily trying to direct every new radio diner to a table with his or her group, seating a group when a table becomes clear, and keeping track of who is sitting with whom.

When a trunking system user presses the talk button on his or her radio, he or she is walking up to the restaurant entrance and requesting a table. Because the radio also transmits the user's group identity, the controller/maitre d' knows whether others from that group are already active on one of the system's channels. If so, the controller/maitre d' directs the new user's radio to the correct channel. If not, the new user is put in a queue for the next available channel.

When the controller has determined what channel the radio user should use, it sends a go-ahead beep and the user can begin transmitting. Think of the beep as the maitre d' saying, "Your table is ready!" The conversation with the maitre d' takes place entirely on the control channel, unheard by any other user.

One important difference between the trunking system and other kinds of communication is that radios using a trunking system can be moved from channel to channel even in the middle of conversations. That makes trunking systems very difficult to monitor without a smart scanner that follows every move the maitre d'/controller makes.

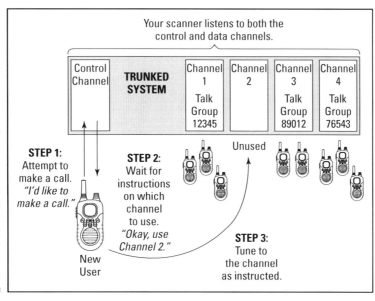

Figure 10-4: A trunking system controller interacts with each user so that channels are used efficiently.

Public service and public safety systems are most likely to use trunking technology; I discuss scanning these systems in Chapter 11. If you seek additional information, you can find an excellent collection of columns by Dan Veeneman on trunked systems from various issues of *Monitoring Times.* Check out www. signalharbor.com/ttt/index.html.

Programming Your Scanner

Programming your scanner isn't an immediate necessity. Most scanners come with some frequencies (and possibly even a few trunked systems) already programmed into them at the time of purchase. All you have to do is drop in a few batteries and away you go! Over time, you can add the frequencies appropriate for your area and delete unwanted channels.

Finding frequencies

Companies and agencies don't publish their radio operations information, so you need to figure out how to find it. By far, the most popular source for frequency information is online:

- ✔ www.radioreference.com lets you set up a free account and get data from its online database. It provides several search formats.

- ✔ www.trunktracker.com is a Web site provided by Uniden. It provides information on trunking systems by state.

- ✔ www.scannermaster.com is a Web site that enables you to order a printed book and CD-ROM that provides public safety, government, and business frequencies.

- ✔ svartifoss2.fcc.gov/reports/index.cfm is an alternate to the FCC ULS Search page. You may find this site easier to use.

- ✔ www.scanningusa.com is home to both the magazine *Scanning USA* and frequency references.

If you're living in Duluth you may not be interested in listening to the transmissions of the local police and fire departments in Colorado Springs. Zeroing in on the information in your local area is crucial. Here's all you need to know to run a search on the RadioReference Web site:

1. **Become a registered user.**

   Visit www.radioreference.com and click the <u>Create an account</u> link. Create a user name, provide your e-mail address, and create a password (you need to enter the password twice).

If you want RadioReference to create a password for you, leave the password fields blank.

2. **Read all the important information and click the Create an Account button. On the screen that appears, click Finish.**

   Within a few minutes you receive an e-mail that explains how to complete the registration process.

3. **Click the link in the e-mail, follow it to the RadioReference Web site, and on the page that appears enter your user name and password.**

   If the first link in the e-mail doesn't work, click the second link and follow the directions to activate your account.

4. **When you're logged on, click the <u>RR Database</u> link.**

   A Web page with a map of the United States on it appears. Below the map is a list called Metro Area Quick-Hops. This list displays the names of major metropolitan areas.

5. **Click an area of the map or one of the cities that you're interested in searching.**

   You can also use the search windows at the bottom of the page to make a very specific search.

   Assuming you search by state, a list of counties appears.

6. **If your county is highlighted (meaning that information is available), click it.**

   A list of trunked systems and some options for viewing lists of licensees appears.

7. **Click the name of the trunked system you want to scan.**

   The information you need appears, including the system type, frequencies, talk group IDs, and more. (See Chapter 11 for full details.)

8. **If you're looking for nontrunked systems, the information is listed below the Trunked System list. Or you can click one of the choices listed under the FCC Data Options heading and search the resulting list for the system you want.**

9. **Keep clicking until you find all the information your heart desires.**

   You can print the info or write it down for future reference.

If the frequencies you find don't work, remember that businesses and agencies occasionally change frequencies and systems. You can either try an alternate database or search the FCC's database directly by using the Universal Licensing System at `wireless.fcc.gov/uls`. Click the Licenses button and then the <u>Advanced License Search</u> link. You may then enter any criteria to narrow your search. When the data appears, click on the call sign to get complete information on the station or system.

Using a PC with your scanner

Although you can certainly program channels individually, even entire systems, this process quickly becomes tedious. If you have a scanner that can interface with a computer, you can use your home PC to make short work of setting up a scanner and sharing frequency information with your friends and associates. Most new scanners can interface with PCs. In fact, if you have a new scanner, your manufacturer probably offers software.

Finding the right software

The manufacturer of your scanner probably offers a software package to work with your scanner. For example, you can download the Uniden Advanced Scanner Director software from the product support page associated with each scanner model. If you have a Uniden scanner, visit www.uniden.com and click the Support link. From there, choose your model from the drop-down list.

Scanner dealers usually offer software packages. Software is also available from independent developers:

- **BuTel Software** (www.butelsoft.com) provides software for Uniden scanners.

- **Computer Aided Technologies** (www.scancat.com) provides software for a wide variety of scanners.

- **StarrSoft** (www.starrsoft.com) provides software for RadioShack scanners.

In my experience, manufacturer software certainly has the necessary functions to get you going — and often for free. However, the third-party developers are a little more competitive, offer better sets of features, upgrade more frequently, and are more focused on user support. The cost of most third-party programs is quite reasonable.

Interfacing the scanner and the computer

Scanners connect to a PC through a serial data link such as RS-232 (PC COM) or a USB port. If you buy a scanner with a PC-compatible interface, the necessary cable should be provided. Check to be sure that the scanner, your computer's serial interface, and the software package you plan on using are compatible.

If your computer doesn't have an RS-232 port (sometimes referred to as a *legacy port*), you may have to purchase a USB-to-RS-232 adapter. Because not all adapters are the same, it's best to check with your scanner dealer or the software provider to see what models are known to work with your system.

Programming scanner memory

The first step to programming your scanner is to organize the frequencies you want to monitor. A group of frequencies used by one agency or city are referred to generically as *systems*. Start by using one of the online or printed references discussed earlier in this chapter. Start small — just create a small system of a few local channels by entering them into the scanner programming software.

Connect the scanner to your PC. Confirm that the software can interact with the scanner by executing a command that requires the software to check the contents of the scanner's memory. If your scanner was sent preprogrammed by the manufacturer, this would be a good time to find out how to *download* the scanner's memory and save it as a backup.

Whenever you start loading new frequency information into your scanner, be sure that the scanner memory is backed up onto the PC and that the batteries are fresh. Alternatively, you should power the scanner by using ac power. If scanner power fails in the middle of reprogramming, you may have to completely clear the scanner memory, losing whatever was stored.

Practice transferring your small frequency list back and forth to the scanner. Add or delete a channel on the scanner and transfer the altered memory to the PC (use a new filename to avoid overwriting your original list). Add frequencies to your original list and transfer the list to the scanner once again. When you feel comfortable with the software, you can start loading your scanner memory in earnest.

Because the scanners available today have hundreds and thousands of channels, this means a lot of typing, doesn't it? Not necessarily, thank goodness! Here are some ideas to save time and effort:

✔ You may be able to copy the memory of a friend's scanner (see the section called "Cloning," later in this chapter) or use your buddy's frequency files if he or she has the same programming software that you have.

✔ If you have joined a scanner club, find out if someone else in the club has files you can use. The club may even have files posted on its Web site. All you have to do is download them.

✔ You can further simplify the process of generating the frequency lists manually with a spreadsheet. If you're careful, you can copy and paste operations from a Web site or frequency list CD.

✔ Users of BuTel software (www.butelsoft.com) find the WebCatcher utility to be a tremendous timesaver. It automatically extracts frequency lists and station IDs from Web sites in preparation for programming!

Organizing frequency files by function makes customizing your scanner for a particular activity much easier. Going hiking? Load the scanner with the package containing state patrol, search-and-rescue, NOAA weather, and CB calling channels. Create a different package for all of your regular activities.

Moving from banked memory to dynamic memory

How you organize your frequency lists will depend to some degree on how the memory in your scanner is organized. Until recently, scanner memory was set up in *banks*, fixed-size amounts of memory that can store some limited number of channels. For example, a four-bank scanner with 100 channels per bank could store 400 channels.

Banked memory is somewhat limited because the scanner can only scan through consecutive channels in one bank before jumping to the next bank. The scanner can't jump back and forth

between banks, nor can unused channels in one bank be added to another bank. Maximizing the scanner's memory often requires some careful planning and eventually leads to difficulties in getting channels to scan in the sequence desired.

Today, more scanners have *dynamic* memory circuits organized in one long block without any banking restrictions. The dynamic memory system eliminates wasted channels and gives you a lot more flexibility when you select and manipulate the scanner's operation.

Cloning

The cloning function is handy because it enables you to duplicate the memory of one scanner in another, often without the use of a PC or external software. For example, you can connect the serial interface cables of two Uniden BC246T scanners directly together by using a *null-modem adapter* (which reroutes data signals so that the scanners can talk to each other) and *gender changer* (which changes male connectors to female connectors).

After you designate one scanner as the master and the other as the slave (sorry to use such barbaric language!), the cloning process proceeds automatically. PC software can also clone *one-to-many*, enabling you to set up several scanners very simply. Check the user's manual of your scanner for complete instructions.

Back up your scanner's memory before doing anything irreversible.

Choosing Scanners and Antennas

If you haven't already purchased a scanner, you may be wondering how to decide between the many fine models available. The key to making the right choice is to cut the problem down to size, one step at a time. Begin by logging on to the Web sites of the major scanner manufacturers and browsing their product lines. In no particular order, here are some manufacturers to check out:

✔ **RadioShack** (www.radioshack.com) has downloadable scanner comparison charts. From the RadioShack home page, click <u>Product Manuals</u> and select Radio Communications from the drop-down menu. The scanner comparison charts are listed below the Documents heading.

✔ **Uniden** (www.uniden.com) is the largest scanner manufacturer in the marketplace. It has a mind-boggling array of products.

✔ **AOR** (www.aorusa.com/main.html) stands for Authority on Radio. AOR is an international company with headquarters in Japan, the U.S., and the U.K.

If you want to go whole hog with your scanning capabilities, check out the *wideband* or *communications receivers* discussed in Chapter 15. These devices provide wider frequency coverage and more signal modes than everyday scanners.

Determining how you expect to use the scanner

If, after comparing all the different scanners on the market, you're still not sure which one you want to buy, evaluate how, when, and where you would like to use the scanner. For example, if you plan on taking it out in the car or on a hike, you can eliminate the *base* models designed to sit on a desk or table.

When in doubt, start with a midrange handheld model and purchase both ac and auto power adapters. A *mobile* unit can make a good desktop model, but will require a 12 V dc power supply.

Dealing with coverage and modes

Not sure what kind of scanner to get? Sit down with a list of bands and decide which you want to monitor. Note whether you want to include coverage of weather channels and broadcast AM, FM, or TV signals. Deciding what you want to listen to can help you determine your scanner needs.

Most scanners can monitor trunked systems, but they may not all be able to handle all the variations (for example, some may be able to handle Motorola I-II-III, but not EDACS or LTR). Check the frequency listings for your area to see what systems are in use. The same goes for digital (P-25 or TETRA) systems. (Trunking systems are discussed in more detail in Chapter 11.)

Evaluating scanner features

Here are the features you should look for in a good scanner. The more basic the need, the closer it is to the top of the list:

✔ **Computer compatibility:** For all but the simplest scanners, PC interface and software support are musts.

✔ **Cloning capabilities:** Cloning is also a very handy feature, particularly in the field, where you may not have access to a PC. Being able to assign an alphanumeric label or *tag* to a channel is very desirable.

✔ **Lots of channels and banks:** In general, a bank should have at least 50 channels and there should be at least 4 banks of memory. Beyond that, more channels and more banks are good. There is no magic combination. If you are just starting, estimate how many channels you think you'll need. Then *triple* that number! A scanner with dynamic memory will be more flexible than one with static memory.

✔ **Decoding capabilities:** Smart receivers can decode CTCSS tones and DCS codes to help you listen to a system without having to know the codes beforehand. Uniden's Close Call feature allows the scanner to find and assign a channel to strong local signals, greatly assisting scanner setup at temporary events or while you're traveling.

Choosing an antenna and accessories

The flexible antenna that comes with most handheld scanners is fine in most situations. To effectively use a handheld scanner in a car, however, you need a mobile antenna that mounts outside the car. The most popular mobile antenna style is a *mag mount,* a *metal whip* with a magnetic base that attaches to the roof or trunk. You can find several examples of mobile antennas on the Grove Enterprises Web site (`www.grove-ent.com/antennas.html`). Here are some good antennas to think about:

✔ **Discone:** At home or in another fixed-location installation, a good scanner antenna is the *discone*, typified by the Diamond D130J (`www.rfparts.com/diamond/d130j.html`). This *omnidirectional* antenna receives signals equally well from all directions. You can easily mount the discone antenna on a lightweight TV mast or a roof drainpipe.

✔ **Beam:** To pull in distant stations, you may need a *beam* antenna that hears best in one direction, such as the Grove Scanner Beam (`www.grove-ent.com/BEAMII.html`).

Beam antennas require a *rotator* to point, just as a TV antenna does. See Chapter 18 for more information about installing antennas; take a look at the Grove Enterprises Web site for antenna illustrations.

Choosing a manufacturer and dealer

Along with the scanner features, you'll require support from the manufacturer as you work with the scanner and questions arise. Also, you may need to contact the manufacturer if the scanner requires repair. Here's a short list of things to consider before buying a scanner:

✔ **Warranty:** The warranty should be at least one year. Extended warranties are a questionable value unless the scanner gets a lot of portable or mobile use.

✔ **Dealers:** Having a dealer near you means better repair and warranty service, plus the opportunity for information and guidance.

✔ **Web site:** You should have an easy method to obtain product specifications, manuals, software upgrades, answers to questions, and important e-mail addresses and phone numbers. The manufacturer's Web site should include all this and be frequently updated.

✔ **Customer Support:** Try calling or e-mailing the customer service department of the scanner manufacturer to see if you can get through to a helpful person. If not, consider buying your scanner from a different manufacturer or dealer.

Chapter 11

Scanning Public Service and Safety Radio Transmissions

In This Chapter

▶ Discovering your local systems

▶ Unlocking the details of trunked systems

▶ Learning the public service jargon

▶ Staying out of the way

*I*n Chapter 10, I cover the basics of scanning and scanners. In this chapter, I show you how to put that information to work. The most common use of a scanner by far is to listen to the operations of the public service and safety agencies. These capabilities aren't limited to police and fire departments; you can listen to transportation, utility, school, hospital, and administration activities of local, state, and federal government workers. Although I focus on public systems in this chapter, you can apply what you ferret out to scanning private radio systems, too.

In this chapter, I show you how to find and transfer the information you need to program your scanner. Building on the introduction in Chapter 10, I take you another step farther into the world of trunked radio systems. Then I tell you the jargon you're likely to hear on the airwaves.

Tracking Down Your Local Government

If you're lucky, your scanner came preprogrammed to scan your local public agencies or a very good friend has a scanner already set up with all the systems you need. Most of us aren't so lucky and even if you are, you'll have to modify or expand the scanner's programming sooner or later. Where do you start?

In Chapter 10, I list several excellent sources of frequency data. Don't forget to check out the support resources offered by your scanner manufacturer, as well.

Acquiring and saving data on your computer

Most scanner enthusiasts like to work from online data sources because they're updated more frequently than CDs or printed references. The most popular of those is RadioReference (www.radioreference.com). All you have to do to start is visit the Web site and register. (I show you how to do this in Chapter 10.)

The rules for accessing data for government and public agencies vary dramatically around the world. If you seek public information, you will follow a process similar to the one I describe in this chapter. Nonpublic data takes more sleuthing to find; you may have to collect it from individuals and clubs or by directly monitoring the airwaves and capturing the active channel information. Be careful and reasonable! Violating scanner or receiver laws could land you in legal hot water!

After you log on to the site, follow these steps to access, copy, and save data:

1. **Click the <u>RR Database</u> link, which appears on the left side of the screen.**

   This link is located below the Radio Features heading.

   The RadioReference data is organized first by state, and then by county. For example, the city of Tacoma, Washington, is in Pierce County.

2. **Find the state and county for which you seek information.**

   From the database page, you can click the state outline and then click the county name. Alternatively, you can scroll down to the Search for Radio Data window, enter the name of the county (say, *Pierce*) in the Search for Agency/County text box, and click the Retrieve button. If you choose this option, a list of all the counties with that name appears. For example, there are five Pierce counties in the U.S.

3. **Select the county you want.**

   A page appears with a map of the county. You probably see several entries for the city you're most interested in. For example, Figure 11-1 shows several listings for Tacoma, Washington. The results in my Tacoma example show two trunked systems and a list of repeater channels.

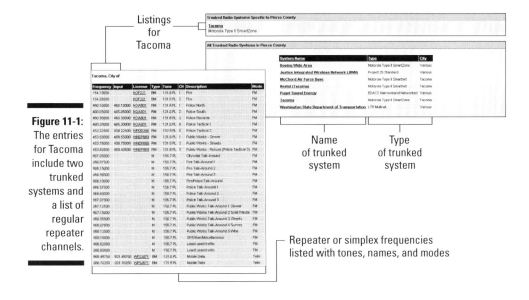

Listings for Tacoma

Figure 11-1:
The entries for Tacoma include two trunked systems and a list of regular repeater channels.

Name of trunked system

Type of trunked system

Repeater or simplex frequencies listed with tones, names, and modes

4. **Open a blank document in a spreadsheet, such as Excel.**

   You can also print the information for transcribing by hand later, but this is a much more efficient option.

5. **Highlight the lines of data in the database and copy them to the clipboard. Paste the data into the blank spreadsheet.**

   The information appears in a table format, although the cells may have strange formatting — don't worry about that.

6. **If the data contains odd characters use the spreadsheet's Find and Replace functions to delete them.**

7. **Highlight and copy the columns of data (start with the Frequency column), pasting them one at a time into your spreadsheet or scanner control software.**

   Figure 11-2 shows the BuTel ARC246 PC software for programming the Uniden BC246T after transferring the police and fire repeater channel data for the city of Tacoma.

8. **Delete any unwanted lines after you're done and then discard the spreadsheet. Edit names of the channels in the scanner programming software and check for mistakes.**

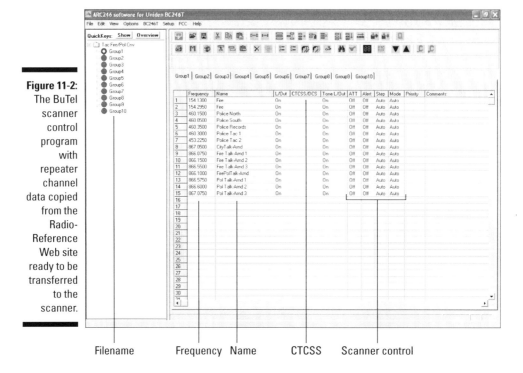

Figure 11-2:
The BuTel scanner control program with repeater channel data copied from the Radio-Reference Web site ready to be transferred to the scanner.

Filename   Frequency Name   CTCSS   Scanner control

9. Save the information on your PC.

Be sure to give it a distinctive name that you will recognize later. These groups of frequencies used by a single organization or company are called *systems* in the scanning world.

Transferring data to your scanner software

There's very little point in accessing, copying, and saving frequency information for public agencies if you don't make this information available to your scanner. Of course, all scanners are different, so refer to your manual for further information. To transfer data to your scanner, follow these steps:

1. Hook up your scanner and transfer the new system data to it.

Follow the instructions in your scanner and software manuals.

Organizing your system data

As you begin to collect system data, you'll soon outgrow the everything-in-one-folder filing strategy. Time for some organization! Start by naming systems consistently, such as `(Tacoma) - Fire` or `Air Traffic Control - (Washington)`. This technique makes finding one fire department among many much easier. Using a dedicated folder for each type or region of data also makes your growing library of information much easier to manage.

2. **Step the scanner through each channel and make sure it understands everything you've programmed it to do.**

3. **If you find any mistakes, edit the information on the PC and retransfer the system data to your scanner.**

4. **Save the system data on your PC once more.**

Scanning Trunked Systems

Acquiring information for trunked systems is almost the same as for regular repeater systems (as described in "Acquiring and saving data on your computer," earlier in this chapter), with one extra step — identifying all the users of the system by the addresses or *talk group identification codes* transmitted by their radios. In Chapter 10, I explain that trunked systems work similarly to a restaurant model, with the trunked system using the identifying code or address to assign users to the proper channel. Your scanner needs to know the addresses and identification, too, so that it can follow the desired conversation as the controller shifts the radios around from channel to channel.

Each of the popular trunked systems handles addresses and identification a little differently, so the following sections offer a simplified description of each. Depending on what scanner or programming software you use, the exact method of entering data about the system varies. You may be asked to select one of the types of systems listed in the following sections, the frequencies the system uses, and the addresses or identification codes of the radios you want to listen to. Your operating manual contains the exact procedure you need to follow. The following sections explain what data you need to program the scanner.

Motorola Type 1

In Motorola Type I systems, identification of the radio begins with a *block* that contains a range of numeric values called *addresses.* Each block contains smaller groups of addresses, called *fleets.* Each fleet contains *subfleets.* An individual radio's ID is listed by its fleet first and its subfleet next, such as 003-28. The number of fleets and subfleets in a block is determined by a *size code.* The overall organization of this type of system is called a *fleetmap,* identified by a numeric *size code,* such as B2. Your scanner needs to know both the fleetmap code and the address/fleet/subfleet information for each type of radio you want to listen to.

For an example of a Motorola Type I system, visit the RadioReference database, click the <u>RR Database</u> link, and on the resulting page, click the <u>Chicago</u> link. On the page that appears, you'll see a list of agency pages and categories. Click the <u>O'Hare International Airport</u> link. The first trunked system listed is the ARINC Type I system. Click the link for this system and scroll down to see the system fleetmap, with its size codes. Farther down the page you can find the individual groups and their addresses listed in the Subfleet column.

Motorola Type 11

Type II systems don't use the block/fleet/subfleet method (refer to the section on Motorola Type I) to identify the radios — they use *talk group identifiers,* which are numeric identification codes assigned to each radio. The talk group ID code is sent with each transmission and tells others, "I am communicating with this group." Each code is a number between 00000 and 99999. Programming a scanner for this type of system is simpler — you just need the system frequencies and the desired talk group codes.

Type II Hybrid systems combine some Type I and Type II addresses. Type II Smartzone systems link together more than one Type II system for wider coverage.

You can see an example of a Type II system by choosing the Type II Chicago O'Hare Public Safety system from the RadioReference database. The system information and frequency lists are similar to Type I systems, but note the difference in the way the radios are organized. The ID for each talk group is listed twice, once in decimal form (DEC) and once as a hexadecimal number (HEX).

Hexadecimal refers to a number system used in computer systems (such as trunking system controllers) that has 16 different digits; 0 to 9 and A, B, C, D, E, F. This is the equivalent of 0 to 15 in the decimal system. Imagine counting if you had eight fingers on each hand! Computer systems often use number systems that are powers of 2, such as hexadecimal or octal, which uses only

0 to 7. Hexadecimal numbers are often preceded by 0x or end with H to distinguish them from decimal numbers.

Motorola Type I and II trunking systems are described in detail in the March 2000 article "Tracking Motorola Systems" in *Monitoring Times* magazine (available on the Web at www.signalharbor.com/ttt/00mar).

EDACS and LTR

Ericsson and General Electric developed the EDACS (Enhanced Digital Access Communications System) trunking system. It is very similar to a Motorola Type I, but it uses agency/fleet/subfleet addressing rather than block/fleet/subfleet addressing. EDACS systems also assign a *logical channel number* (LCN) to each channel frequency and use this LCN to redirect the radios. Similar to the Motorola Type II systems, it's necessary to program the scanner with both an LCN and the system channel frequencies.

LTR (*logical trunked radio*) systems exchange control information as subaudible tones right along with the voice information — there is no separate control channel. The talk group IDs on LTR systems look similar to EDACS and Motorola Type I; they're constructed from the group's *home repeater* followed by a group ID.

For examples of EDACS and LTR systems, visit the RadioReference database by clicking the RR Database link. Click a state on the map, scroll down the resulting page, and click the link to see all trunked systems for that state. For example, if you click the state of Illinois, scroll down the resulting page and look for a link that says Click Here to View All Trunking Systems in Illinois. You should get a good array of all the trunked systems available. For example, in Illinois, the Lake County Public Safety System is an EDACS system and ESP Leasing uses an LTR system. Articles explaining EDACS and LTR are available at www.signalharbor.com/ttt/index.html.

Setting Up a Trunked System on Your Scanner

Yes, trunking systems can be complicated, and your first will be the most difficult. (If you haven't joined a scanner club yet, you need one now more than ever!) My recommendation is to copy the scanner setup for a similar system already working (borrow it from a friend) and adapt the setup to your system based on the information in the database. Here are three primary guiding resources:

✔ **The manual:** Your scanner manual probably includes a blank template for you to fill out with system information; it probably offers step-by-step instructions, too.

✔ **The software:** If you're using scanner-programming software, templates are probably already set up to guide you in obtaining and entering data.

✔ **The Web site:** The manufacturer of your scanner is likely to have step-by-step instructions for programming trunked systems on its Web site.

Here are some steps to help you find, transfer, and save information for a trunked system onto your scanner:

1. **Find and click a trunked system about which you want to access data.**

   For example, if you want to find a trunked system in Tacoma, you're in luck. The trunked system listings are always listed just above the regular repeater channels for a region. Tacoma has two trunked systems, one labeled Nextel (Tacoma), and the other called simply Tacoma.

   Figure 11-3 shows the page that appears when you click the Tacoma option. (This system just so happens to be used by the police and fire departments in the city of Tacoma.)

2. **Determine which type of system you're dealing with.**

   For example, the figure shows three basic sets of information: system description, system frequencies, and system talk group IDs. These describe a Motorola Type II Smartzone 800 MHz trunked system.

3. **Identify the system type to your scanner programming software.**

   For example, in your software, you might see a listing for Motorola Type II 800 MHz Standard. This is a pretty good match for the system that's used in Tacoma, so select this listing.

   If you're programming the scanner to identify a Type I system, you should also have to enter the fleetmap information.

4. **Copy the channel frequencies.**

   Follow the technique listed in "Transferring data to your scanner software," earlier in this chapter, to transfer the frequencies from the Web page to your programming software.

   If you are working with an EDACS system, be sure to copy the LCN information, too.

   At this point, you have configured the system and are ready to list the talk groups you want to listen to.

5. **Copy the talk group IDs.**

   Transfer the talk group ID information (otherwise known as *addresses* in Type I, EDACS, and LTR systems). Edit the names so that they make sense on your scanner's display.

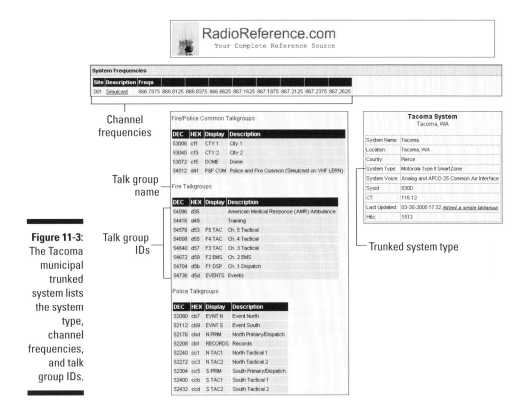

Figure 11-3:
The Tacoma municipal trunked system lists the system type, channel frequencies, and talk group IDs.

6. **Transfer the information into your scanner.**

Congratulations — you've just set up a trunked system! Press the Scan button and start listening. Your scanner will automatically find the control channel (or decode it directly on LTR systems). Me? I can't wait to listen to those garbage trucks on pickup day!

Cracking Codes and Learning Lingo

As soon as you start scanning public service and safety frequencies, you'll find yourself hearing some peculiar radio lingo. It won't sound like normal speech, nor should it! The language of safety and law enforcement must be clear, unambiguous, efficient, and precise. Officers and public safety workers are continually trained in the proper use of their radios. When time is of the essence and someone's fate may be hanging in the balance, there's no time for sloppy radio technique! Here are some common terms you'll run across:

- **Dispatch:** The central coordinating station or center. Dispatchers are the persons that operate the dispatch center.

- **ETA:** Estimated time of arrival.

- **Hazmat:** *Haz*ardous *mat*erials, generally referring to a spill or some other situation where hazardous materials are involved.

- **Mutual aid or assistance:** Personnel of more than one department responding to an incident (such as police and paramedics). Large systems and regions generally have at least one mutual aid talk group or channel.

- **Operations:** Communications to coordinate the day-to-day activities of a department or fleet.

- **Tactical:** Communications that coordinate the response to an incident. Channels or talk groups are often set aside just for tactical communications.

- **Talkaround:** An intercom channel or talk group everyone can use.

10-4, I am 10-23, waiting for a 10-28. Then I'll be 10-98.

Long-time scanners and fans of the old police shows on TV are familiar with 10-codes. (If so, say, "10-4.") Although the use of codes has declined in favor of plain talk, you can still hear a lot of verbal shorthand.

Variations on 10-codes are common; you can see abbreviations and numeric codes listed at `www.cobras.org/police.htm` and on the California Highway Patrol site, `cad.chp.ca.gov/body_glossary.htm`. Don't spend too much time memorizing codes, though, or you'll miss all the action!

The 10-codes were developed by the Association of Public Safety Communications (`www.apcointl.org`) years ago in an effort to standardize common communications. The complete list is available on the Scanner Club of Ohio's Web site at `www.aosc.org/apco.html`. Here are some of the more common codes:

- 10-4: Acknowledgement (not an agreement or acceptance).

- 10-6: Busy, stand by.

- 10-20: Location, as in, "My 10-20 is First and King."

- 10-23: Arrived at scene.

- 10-28: Vehicle Registration, as in, "The vehicle's 10-28 is Mr. John Doe."

- 10-40: Back up (reinforcement).

- 10-98: Finished with last assignment.

Incidents that require more intensive response levels are now managed by the *National Incident Management System* or *NIMS*. The radio activities and titles used during such emergencies make a whole lot more sense if you're familiar with NIMS. You can read up on NIMS and even take online courses at `www.fema.org/nims`.

When you're listening to police channels, you'll also hear references to *vehicle codes*, which refer to specific traffic laws. Each state and municipality has its own set of vehicle codes.

Helping, Not Hindering

Rushing off to a fire or accident or crime scene based on what you hear on your scanner is not a good idea. Don't become part of the problem. In efforts to get a look at the action, you could clog the roadways or get in the way. At worst, you place yourself in potential danger, diverting attention from the original emergency — endangering multiple lives. You can be cited for interfering with public safety operations if you get in the way, so don't be an obnoxious and unwelcome wannabe. If you want to help out, you can find plenty of opportunities.

If you're motivated to help, every police and fire department has volunteer and auxiliary organizations that welcome your assistance. Here are some ideas to put your energy to work:

- **CERT, Citizen's Emergency Response Teams** (`training.fema.gov/ EMIWeb/CERT`)**:** CERT consists of citizens trained to respond in their neighborhoods, usually coordinated by one of the local public safety agencies.

- **Amateur Radio** (`www.arrl.org`)**:** The most powerful of all citizen radio services (see Chapter 9) includes numerous emergency and disaster response teams.

- **REACT, Radio Emergency Associated Communications Teams** (`www. reactintl.org`)**:** Originally associated with Citizens Band, REACT now incorporates all kinds of radio users.

- **SKYWARN** (`www.skywarn.org`)**:** A volunteer organization of severe weather watchers that report local conditions to the National Weather Service.

- **Fire and Police Department Auxiliary:** Check with your local departments for opportunities to be trained in a variety of useful ways.

Chapter 12

Radio Aloft: Aviation Radio Transmissions

*N*ext to the public safety channels, where mayhem and miscreants are never far away, the aviation channels provide a continuous source of interesting listening material. If you live near a busy metropolitan airport, the action is nonstop! Off-duty pilots monitor aviation channels like drivers monitor the traffic reports. Pilots in training monitor to master good operating procedure. Civilians just love the crisp back and forth of the airways.

In this chapter, I take you on a quick fueling stop at the frequency listings to find out where to listen and who will be transmitting. Next, you can taxi out to the runway, taking in some information about airport operations. Having taken off, I cover what you will hear from pilots when you cruise the airwaves. Finally, you can check off some waypoints of aviation jargon.

Activity on the Aviation Bands

There are a number of bands where aviation communications are present, but the lion's share of interesting conversations take place on what is called the *Aircraft* or *Air Band* between 118 and 137 MHz. These are the frequencies where pilots talk to each other and to air traffic controllers. Airport operations are also conducted here. These are *civil aviation* frequencies. (I cover military aviation in Chapter 13.) Complete information about the rules and regulations of the FCC's *Aviation Service* is available at `wireless.fcc.gov/aviation`.

Aviation communications or *traffic* channels begin at 118.000 MHz and are spaced every 0.025 MHz (or 25 kHz) through 136.975 MHz. Regional air traffic control centers also use frequencies in the UHF air band above 250 MHz. Unlike public safety and most other VHF and UHF services, the aviation standard is amplitude modulation (AM) signals. This is because AM performs better when signals are weak, a common issue, considering the long distances over which aircraft need to communicate. Aviation communications are also *simplex;* planes don't need repeaters or trunking systems to extend their range because they're already operating at heights greater than any repeater.

If you want to listen to aircraft channels, be sure to confirm that your scanner or receiver can receive AM signals on the *aircraft band.* Don't get confused — AM *broadcast* signals are in a completely different part of the radio spectrum.

Finding Frequencies

You can look up the frequencies in use at an airport by visiting a Web site such as AirNav (`www.airnav.com`). AirNav enables you to view a complete set of fascinating information about any airport in the United States (and its possessions). For readers outside the U.S., try World Aero Data (`worldaerodata.com`), which lists airports from around the world.

Each airport has an alphanumeric *identifier.* For example, Seattle-Tacoma International Airport (often called Sea-Tac for short) uses the identifier SEA. Smaller airports get more cryptic identifiers such as 0G0 (that's zero-G-zero) for the North Buffalo Suburban Airport in western New York state. Planes also have an identification number, often referred to as the *tail number,* because that's where the number is usually displayed.

The international aviation emergency frequency is 121.500 MHz. Pilots in distress use this frequency and *ELTs* (Emergency Locator Transmitters) do too. (ELTs activate when an airplane crashes or if the pilot turns it on, acting as a beacon for rescuers.) Survival craft and some ELTs also transmit on 243.000 MHz. These are good frequencies to have on any scanner's priority channel list.

Here are several frequencies that have common uses in most areas of the United States and Canada:

- **122.200 MHz:** Federal Aviation Administration (FAA) or Transport Canada flight service
- **122.750, 123.025 MHz:** air-to-air coordination
- **122.850, 122.900 MHz:** Multicom (see the following section on airport interactions)
- **122.950 MHz:** Unicom (see the following section on airport interactions)
- **123.100 MHz**: Civil Air Patrol Search and Rescue (U.S.)

During wildfire season, the frequencies 118.925 MHz, 118.950 MHz, and 119.50 MHz are used for airborne firefighting, otherwise known as smoke-jumpers. In calmer periods, listen at 122.250 MHz to hear balloonists talking to each other.

Ground Control to Major Tom: Airport Operations

Airports often list many frequencies, so knowing which ones to monitor can be tricky. The busiest and most interesting frequencies illustrated in Figure 12-1 are listed as

- **Tower:** The primary frequency for pilots to coordinate with the airport control tower at controlled airports where operators manage all landings and takeoffs.

   Controlled airports are airports with a control tower that coordinates landings and takeoffs with pilots. *Uncontrolled* airports are small, regional airports that have no control tower.

- **Ground Control:** The frequency to which pilots tune for instructions when they are on the ground at controlled airports.

- **Unicom:** An intercom frequency mostly used by small aircraft to coordinate with other pilots and airport control towers. Pilots use this frequency to inquire about refueling and other services.

- **Multicom:** Another intercom frequency primarily used near small, uncontrolled airports. Pilots use the multicom frequency to coordinate landings and takeoffs in the absence of a ground control station.

- **ATIS (Automatic Terminal Information Service):** A continuously broadcast summary of weather and operating conditions at the airport.

After you have located the main frequencies for a particular airport, you may want to add the *approach* and *departure* frequencies. At larger airports, activity on various runways and approach paths are managed on separate frequencies. Pilots are directed to tune to these frequencies when they check in with the control tower or ground control.

So many different operations are going on at large airports that you probably won't want to have your scanner scan all of them — you'd miss most of the good stuff between the tower and pilots! Use the *lock-out* feature of your scanner (described in Chapter 10) to skip the channels of lesser interest; then tune to those lock-out channels when something interesting is happening.

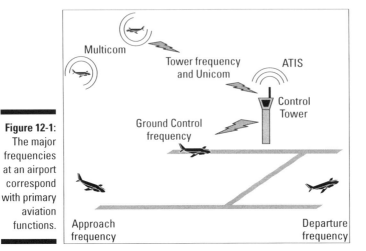

Figure 12-1:
The major frequencies at an airport correspond with primary aviation functions.

Listening to Air Traffic Between Airports

In addition to listening to traffic at airports, you can also hear what's being said when planes are en route to their destinations. When pilots are airborne in wide-open spaces, they work with regional centers called Air Route Traffic Control Centers (ARTCC). The primary task of these centers is to make sure that commercial planes have the support they need while they are en route to their destinations. You can hear these communications if you know where to listen for them.

There are 20 ARTCC centers in the continental U.S., plus one in Anchorage, Alaska, and one on the island of Guam. Each center coordinates a region that contains a number of ground stations that are responsible for smaller areas called *sectors*. The airspace above the sectors is divided into low-level (meaning low-altitude) and high-level (high-altitude) sectors, as well. Figure 12-2 shows the Eastern and Western U.S. ARTCC regions and lists the name and three-letter identifier for each. European air space is similarly managed by Eurocontrol, the European Organisation for the Safety of Air Navigation (www.eurocontrol.int).

To find your ARTCC region, browse to the Web site at www.milaircomms.com/artcc_frequencies.html. Click a region to load a map with a list of low-level and high-level frequencies. Unless you live within 10 miles of the ground station, you probably won't hear the ground side of conversations, but you can hear airborne planes for quite a distance.

Receiving messages from the sky with ACARS

ACARS stands for *Aircraft Communication Addressing and Reporting System.* Commercial aircraft send digitally encoded ACARS messages at 131.550 MHz to ground stations located around the world. (This is the worldwide channel. There are also regional and alternate channels.) The messages contain a lot of information about aircraft status, position, and weather. The messages are sent to ground stations maintained by ARINC (an aerospace company) and relayed to the company that operates the aircraft. If you have an aircraft band receiver and a computer with a sound card, you can use software to decode these messages for yourself. They sound like a rasp on a voice-only receiver. (Chapter 20 has more information about using a computer and radio together.) For more information on how ACARS works, frequencies for ACARS messages, and software to decode the messages, log on to www.acarsonline.co.uk.

Figure 12-2: The 20 ARTCC centers in the U.S. with the three-letter identifiers for each one. Each center has several ground stations.

West Region

U.S. ARTCC Centers

East Region

Albuquerque ZAB
Chicago ZAU
Denver ZDV
Fort Worth ZFW
Houston ZHU
Kansas City ZKC
Los Angeles ZLA
Minneapolis ZMP
Oakland ZOA
Salt Lake City ZLC
Seattle ZSE

Atlanta ZTL
Boston ZBW
Cleveland ZOB
Indianapolis ZID
Jacksonville ZJX

Memphis ZME
Miami ZMA
New York ZNY
Washington, D.C. ZDC

If you have a wideband or communications receiver, listen up. Pilots use VHF and UHF frequencies when they fly over land; when they fly longer routes over oceans, they use a number of lower frequencies. These are in the HF part of the radio spectrum below 30 MHz. You can view a list of these *MWARA (Major World Air Route Area)* frequencies at www.grove-ent.com/mtmwara.html. See Chapter 15 for more information on receiving these frequencies.

Strangling Your Parrot: Aviation Jargon

The phrase *strangle your parrot* is a good example of the obscure jargon pilots like to use. It stems from World War II, when the first onboard systems to automatically *identify friend from foe* (IFF) were created. The IFF systems were notorious for making noise when they transmitted codes. (Guess what the chirping and squawking sounded like!) Remember, anyone with radio equipment can hear communications, even the enemy. So, to avoid tipping off the enemy that they had IFF systems, American pilots referred to the IFF system as *the parrot*. When a tower operator wanted the pilots to shut off the IFF system, it was time to strangle the parrot! Awk!

Here are just a few of the most common terms and acronyms used on the aviation frequencies:

- **ETA:** Estimated time of arrival
- **ETE:** Estimated time route
- **FBO:** Fixed-base operator, a company that performs aviation and maintenance at the airport
- **IFR:** Instrument flight rules, or flying according to instrumentation
- **MOA:** Military Operations Area such as an air force base or training area
- **NOTAM:** Notice to Airmen, a bulletin regarding the status of an aviation facility or navigational aid
- **Roger:** Received and understood message
- **Squawk:** Send identifying codes
- **Touch-and-go:** Practice landing briefly, and then take off again without stopping
- **VFR:** Visual flight rules, or flying according to visual observation
- **Wilco:** Will comply

Acronyms are frequently used in aviation because the pilots and controllers need to be brief, yet precisely and thoroughly understood. Reliability is crucial, so acronyms and abbreviations provide the perfect solution.

To really understand what you're hearing, *Say Again, Please,* 2nd edition, by Bob Gardner, published by Aviation Supplies & Academics, is a great guide to radio communications. (Go to `www.asa2fly.com` and enter ***say again please*** in the search window.) The book presumes a small amount of aviation background, but is very useful as a reference. Gil Cole's introduction to aviation scanning at `www.qsl.net/n4jri/air_gen.htm` is also a good resource.

Chapter 13

Radios in Uniform: Government Radio Transmissions

. . . .

In This Chapter

▶ Scanning military operations

▶ Listening in on the government

▶ Using common sense when you scan

. . . .

*I*n Chapters 11 and 12, I show you how to scan public agencies and aviation frequencies. In this chapter, I cover the third major source of scanning interest — federal government and military operations. Here you find nearly a parallel universe; rather than police and fire departments, you can listen in on the armed services. You can also listen in on military aviation activity. Workers at federal agencies, such as the Department of Energy, use communications services just as municipalities and states do; you can listen to their conversations, too! The challenge to listening in on government operations is that military and federal radio users aren't required to register their systems with the FCC, so there's no public database of frequencies.

Because the operating procedures are in many cases similar to those in the commercial and regional government sectors, you should refer to Chapters 11 and 12 for more information. This chapter focuses on discovering which agencies are active. It also helps you locate the frequencies for these agencies. The chapter concludes with some insights on how to conduct your scanning in the post-9/11 world.

Scanning the Military

Finding military operations frequencies is tougher than finding public and commercial frequencies. After all, for obvious reasons, the military doesn't publish directories, so the curious scanner user needs to be smart, dig deeper, and be willing to share information with other responsible scanners. The military's understandable tendency towards secrecy is counterbalanced by a healthy and completely legal community of listeners that glean information from Web sites, public documents, and by just plain listening. After all, radio signals don't stop at the edge of the military base!

Several Web sites are devoted to cataloging and updating military frequencies around the world. The U.S., Canada, and the United Kingdom have the most vigorous listeners, judging by the amount of information available.

If military scanning is important to you, bookmark a few of these sites in your browser — the active frequencies change regularly, so you're likely to visit the sites often. Here are just a few sites you should check out. Use the search term *military frequencies* for more information and to locate other scanner hobbyists that like to lurk these frequencies:

- ✔ **Monitoring Times** (www.monitoringtimes.com)**:** This site offers a fairly complete presentation of military frequencies. Find it by visiting www.monitoringtimes.com/html/mtMilVHF.html.

- ✔ **Freq of Nature** (www.freqofnature.com)**:** This site is an excellent general reference.

- ✔ **Military Air Commmunications** (www.milaircomms.com)**:** This site specializes in military aviation frequencies for U.S. forces.

    Readers outside the United States can apply the same general approaches used in this chapter to searching for frequencies in their own countries.

One of the most inclusive one-stop-shopping references for military frequencies is the Grove Military Frequency Directory CD-ROM (visit www.grove-ent.com/SFT31.html). Although military frequencies tend to change more than those of public systems, this exhaustive reference gets you up to speed and provides names for speeding up your online frequency searches.

There are several bands set aside for the exclusive use of the federal government and the military. These are the bands you should focus on scanning. In Chapter 10, I provide a table with all the basic radio bands, but here are a few that you should know:

137.0 to 143.9875 MHz	Military Aviation Band
162.0 to 174.0 MHz	Federal Government Band
225.0 to 400.0 MHz	Military Aviation Band
406.0 to 419.9875 MHz	Federal Government Band

The military is a heavy user of HF (below 30 MHz) radio. Most scanners won't tune below 25 MHz, so to listen to HF channels, you need a wideband or communications receiver. The military uses HF because operations conducted during exercises and hostilities need to operate with a minimum of supporting systems, communicating as directly as possible — the strength of HF signals is perfect for such activities.

Finding military facilities

There are many types of military facilities — forts, bases, airfields, harbors, depots, and so on. In fact, the Puget Sound of Washington state has one of just about every type. The Madigan Army Hospital near Tacoma is a good example. This hospital handles injuries from nearby Fort Lewis and McChord Air Force Base; Madigan also handles military *medevac* (medical evacuation). It's a busy place and a prime candidate for a set of scanner frequencies. Here are the frequencies that Madigan uses for various areas of the hospital:

Medevac	40.55, 47.50
Fire	165.0875, 173.4125, 173.4875
Ambulance	139.375
EOD (Explosive Ordnance Disposal)	49.70, 49.80
Security	36.50, 149.525, 149.875, 150.435, 150.725, 407.375, 412.875, 413.075, 413.425
CID (Criminal Investigative Division)	32.80, 407.375, 413.4125, 413.425
Medic Pager	407.325

These military frequencies are in federal government allocations that are separate from the similar public safety VHF and UHF bands used by civilian services, such as police and firefighting resources. The military versions of civilian services have a heavier focus on security.

I found these frequencies by entering *madigan army hospital frequencies* into an Internet search engine. The easiest way to find frequencies in your region is to search by the name of regional facilities, such as bases, depots, and so on. If you are searching from a master list, start by searching by state and then search within those results by facility name.

Finding armed forces facilities

Although the military relies heavily on HF for *tactical communications* (communications that coordinate the operations of forces during exercises or maneuvers), VHF and UHF radio frequencies are used to conduct day-to-day operations. The trend at all major military facilities is to phase out repeater systems in favor of the trunked systems. This trend is consistent with the tendency of municipal government and public services, as well as larger companies, to use trunked systems. I discuss trunking in Chapters 10 and 11.

Working around talk group IDs

Talk group IDs for armed forces facilities are fairly difficult to obtain, but the latest scanners can capture and decode them directly from the radio transmissions. If you're a persistent listener, you can then build up a fairly complete system with this information.

Navy ships also use trunked systems. Aircraft carriers are so large that they require a system as big as the ones used by some small towns. The USS George Washington uses a 10-channel Ericsson EDACS system, for example. Because the military does not, for obvious reasons, publish the talk group IDs, you need a scanner that can automatically analyze and track the trunking system operation. Uniden scanners that incorporate TrunkTracker technology are very valuable when you are trying to monitor a system without a complete set of information about it.

For example, Fort Lewis is a large Army base near Tacoma. Most operations take place on a Motorola Type II trunked system, using channel frequencies of 406.950, 407.250, 408.550, 409.150, 409.350, and 410.150 MHz. This information is available on the Scanning USA Web site `www.bearcat1.com/fleetwa.htm`.

Monitoring military aviation communications

Military aviation frequencies are by far the richest set of military frequencies your scanner will pick up. Military aviation facilities are present in every part of the world, and military airfields are busy places, day and night. The military aviation bands are at 137 to 144 and 225 to 400 MHz (see Chapter 10 for a more complete table of bands).

Military aviation communication is similar to civil aviation; flight communications are organized around the airfield and en route traffic control systems; operating procedures are also similar, and most signals are transmitted using AM frequencies.

 Although you may recognize much of the language right away, for definitions of the many abbreviations, acronyms, and jargon used in military aviation, try the glossary at `www.danshistory.com/glossary.shtml`.

Military aviation frequencies are available at numerous online sources. Simply enter the name of an air force base, along with the word *frequency* into an Internet search engine. Look for frequencies for the following items at most airfields (the frequencies listed are for McChord):

Arrive/Depart	126.5, 391.9 MHz
Tower	124.8, 236.6, 259.3 MHz
Ground	121.65, 275.8 MHz
Security	165.1625, 413.125, 413.825 MHz
MAC (Military Airlift Command) Operations	407.425, 413.125 MHz
Weather	342.5 MHz
Fire Dept/Crash	173.5875 MHz
Officers Net	165.1125 MHz
Motor Pool	163.5875 MHz
Fueling	148.095 MHz
Maintenance	149.265 MHz
Crew Alerts	413.45 MHz

Military air shows, demonstrations, and display teams, such as the U.S. Air Force's Thunderbirds, the U.S. Navy's Blue Angels, Canada's Snowbirds, and the U.K.'s Red Devils use their own special frequencies. Enter the team name and the term *frequencies* into an Internet search engine to find the list. Scanning at air shows is covered in Chapter 14.

You can make a lifetime career of scanning military aviation frequencies, and many people do. By its nature, military aviation is not a public system; it's meant to stay cloaked in secrecy. This secrecy is part of the allure for many hobbyists. However, for the uninitiated, the details remain frustratingly out of reach. The bottom line is that some information is only available to the serious student — and getting the skinny on how to get information like this goes well beyond the introductory nature of this book. Don't give up! Many frequencies used by military aircraft and facilities are well known. You can access lists of frequencies at Web sites such as Freq of Nature (www.freqofnature.com). This site lists many military and government frequencies.

Accessing Civilian Agencies

Along with the military, the federal government has many civilian agencies, such as FEMA and NASA, which use radio systems, including trunked systems. The trick to zeroing in on useful frequencies is to search the Internet by using the name of the agency (search for *fema frequencies*, for example, to access frequencies for the Federal Emergency Management Association). A good reference source is the Grove Enterprises Federal Frequency Directory (www.grove-ent.com/SFT32.html), published on CD.

As an example, the following frequencies are used nationwide by FEMA during emergencies. (FEMA uses additional frequencies, including frequencies that coordinate with local, state, and military operations.)

TAC (Tactical operations)	142.3500, 142.4250, 142.2300, 143.0000, 142.9750, 142.3750 MHz
General Operations	141.8750, 138.2250, 139.8250, 141.7250, 167.9750, 167.9250 MHz
Mobile Units	164.8625, 165.6625 MHz

For general information on frequencies for a specific federal agency, such as the Departments of Commerce, Transportation, or the Interior, use the Freq of Nature site (www.freqofnature.com). Click the <u>Frequencies</u> link, and then click the <u>United States</u> link. From there, you're free to click the name or group of agencies of interest.

To find local frequencies used by these agencies, I recommend that you start your searches on general frequency reference sites, such as Monitoring Times (www.monitoringtimes.com). Then glean additional local or regional frequencies for the agency of interest by adding your city or state name to the search criteria. If an emergency is actually ongoing, use your scanner's search or strong-signal acquisition features (such as Uniden's Close Call feature) to find active frequencies.

Although most NASA activity is concentrated in the Cape Canaveral, Florida, and Houston, Texas, areas, after a satellite or Space Shuttle is launched, there's a good chance that you will be able listen in as it passes over your region. The International Space Station is often audible on several frequencies, too.

Step Away from the Radio: Following the Rules of Sensible Scanning

Government and military radio users have always been aware of and accepted that their transmissions are part of the public domain and are likely to be monitored. That's the nature of radio signals. Some countries tightly restrict or forbid monitoring, so be sure that you understand the laws wherever you are. Even in the wide-open environment of the United States, however, you can still attract a lot of unwelcome attention by taking foolish risks.

Since September, 2001, all federal and military agencies have stepped up their security measures dramatically. Access to frequency and operations information has been curtailed. Agencies such as the FBI are paying much closer attention to radio activities that could be construed as surveillance by hostile groups or individuals. This is something you should consider, particularly when using a portable scanner in your car or elsewhere.

Yes, it's still a free country, and you are entitled to listen to (almost) anything that your radio can hear. However, if you head to the perimeter of a sensitive facility of any sort, a national border, airport, or other transportation hub, and start operating a scanner, you're far more likely to be visited by security personnel than in previous years and you run the risk of being temporarily detained. This is a major hassle for you and wastes the time of security personnel. If you don't want to take that risk, be reasonable and back off. In general, if you stay away from the boundaries of secure areas, don't loiter or park near building and base entries, and aren't acting sneaky or furtive, you'll be okay.

Chapter 14

Radio in Action: Recreational Radio Transmissions

*B*esides its use to listen in on law enforcement and military activities, radio technology is also used at recreation and entertainment events, sometimes by the participants. With your trusty scanner in hand, you can keep up with fast-changing happenings on the field and track. At concerts and shows, you can listen in and find out why the second band hasn't started yet!

I start out this chapter with one of the hottest uses for scanners — monitoring automobile races. After a lap or two, I switch gears to show you how to apply all your aviation know-how (see Chapters 12 and 13) to monitoring air shows and fly-ins. Then it's over to the stadium to take in some signals at the ballgame. Or maybe you can keep an ear on how the marathon is going. Weekend festivals and evening concerts are more radio intensive than you may have thought, and this chapter shows you how to have a whole new kind of fun with these events. Why not take a scanner along? As you can see, there's a lot of ground to cover.

Now, listen here

A radio wave is a radio wave — and the U.S. operates on the "everything is permitted unless it's forbidden" principle. To put the policy succinctly, if *you* radiate it, *I* have a right to receive it unless the law specifically prohibits me from doing so (it doesn't matter whether you intend your transmission to be private). It's okay for theater owners to request that all receivers be turned off on their property, for example, but the transmissions from inside may be received outside on the street.

Users of wireless devices — microphones, cordless phones, and baby monitors — are often unaware that they are essentially broadcasting transmissions that can be received for quite a distance. They may have an unreasonable expectation of privacy. The FCC is quite clear; the airwaves are public except for three exceptions: mobile phone conversations; paging signals; and scrambled or encrypted transmissions. If you don't want your conversations overheard, encrypt or scramble or your voice . . . or don't use wireless devices.

Taking a Scanner to the Races

The popularity of using scanners to monitor the action at theracetrack has accelerated greatly over the past several years. In fact, more people are discovering the world of scanning through racing than for any other reason! If you're one of those new scanner users, welcome to the wide, wonderful world of radio!

Scanning at the racetrack is becoming so popular that manufacturers are starting to produce radios specially designed for use at theracetrack. A couple of well-known examples are the Uniden SC230 and the RadioShack Pro-99. Each of these radios comes preprogrammed with a complete set of frequencies for the cars and their crews. They're even set up so that the frequencies are organized by car number; that way, you can easily pick and choose which drivers and pit crews you want to listen to. Ready? Go!

Discovering what you can hear

Although you certainly want to hear what the drivers have to say, there's a lot more conversation at your disposal. The crews discuss a lot of technical information that puts the race in a whole new light. Track officials say plenty to each other about how the race is proceeding; they discuss infractions and all sorts of interesting stuff you'll never hear on the broadcast media. Eavesdropping on the announcers leads to some *very* interesting insights — when they talk off camera, you might hear them talk about who they *really* think will win the race.

Race radio transmissions use channels licensed by the FCC. Most frequencies are in the UHF band between 450 and 470 MHz, but use of 800 MHz frequencies is becoming more common.

Listening to drivers and crews

Each team in a series of races (such as the Nextel Cup or Craftsman Truck series) is headed by a driver and assigned a car number for the entire series of races. Each team is referred to by the driver's name, such as Dale Earnhardt, Jr., who drove car #8 in the NASCAR Nextel Cup series in 2005. (If you're new to racing, NASCAR stands for National Association for Stock Car Auto Racing.)

Each team uses the same frequency (or one of a small group of frequencies) throughout the racing season. For example, in 2005, Dale's primary frequency was 464.9500 MHz; his secondary frequency was 463.7250 MHz; his alternates were 453.7250 MHz and 451.9000 MHz.

The driver of the car uses his radio to communicate with the crew chief and sometimes with crew members. They'll discuss how the car is running, how the other drivers are doing, and where they are, whether to come in to the pit for fuel or service, and other important topics. Consider the fact that drivers are averaging speeds of 200 mph while they chat and you may well be impressed. I wish drivers using their cellphones were as talented!

Listening to officiators, emergency response, and track managers

Without radios, managing even a minor race would be impossible. Track officials and support staff use radios to keep the event under control from the stands to the parking lots. Remember that the larger tracks cover almost a square mile. Medical and fire crews are always standing by, ready to respond to the inevitable crashes. No matter where you are in the stands, you'll be able to quickly determine what cars are involved and whether anyone is hurt. Any racing fan knows well the drama of the ambulance and fire trucks rushing to the scene and the anxious moments before a driver is helped from the mangled vehicle. The response, of course, is coordinated by radio with some communications on track channels and some on public safety channels.

Here are some frequencies used by NASCAR officials during the 2005 racing season (remember that they might change at any time):

- 451.1750 MHz
- 451.2750 MHz
- 451.4250 MHz
- 451.4500 MHz
- 451.5750 MHz
- 461.2000 MHz

  ✓ 463.6250 MHz

  ✓ 464.6000 MHz

  ✓ 464.7750 MHz

Listening to the broadcast crews

Here's an unexpected bonus — with a scanner, you get to listen directly to the network announcers as they call the race! Instead of running long cables to the satellite transmitter, the announcer's voices are mixed together into a single audio channel in the broadcast booth and sent to the transmitter by radio. Here are some broadcast stations for you to tune in to when you get to the track:

  ✓ **Motor Racing Network (MRN):** 454.000 MHz

  ✓ **Performance Racing Network (PRN):** 454.000 MHz

  ✓ **NBC:** 450.3875 MHz

  ✓ **FOX:** 450.525 MHz

If you don't hear broadcasters on these frequencies, have your scanner search for active channels between 450 and 460 MHz.

A frequent visitor to racetracks around the country is the Goodyear blimp. Enter the search term *goodyear blimp frequencies* into an Internet search engine to find a whole set of channels to scan. Now, if only your scanner could just show the video, too!

Getting the inside track on frequencies

If you buy a scanner made for racing, it comes preprogrammed with a set of frequencies; it may even offer a subscription to a service that provides frequency and other information. Preprogrammed frequencies work for quite a few of the teams and other track frequencies. However, teams sometimes change frequencies and new teams are being added. The following sections show you how to find them.

Locating team frequencies by using the Internet

Start by becoming familiar with one of the racing Web sites that provides frequency information. A good mix of free and subscription sites is available. Here's a small sample:

  ✓ **Racing Electronics** (www.racingelectronics.com): A scanner and accessories dealer with an extensive frequency subscription service

  ✓ **Race Scanners** (www.racescanners.com): A subscription site that also sells scanners and accessories.

> ✔ **The Paddock** (`motorsports.thepaddock.com`): A free site with several sets of frequencies and a good introductory article on race scanning.
>
> ✔ **National Radio Data** (`www.nationalradiodata.com`): A subscription site with some free information.
>
> ✔ **Northwest Racing Enterprises** (`www.northwestracing.com`): A racing parts and materials supplier with a good set of free links and information.

If you only follow one series or just want to check frequencies before you leave for the track, try an online search using just the name of your series, such as *winston cup frequencies.*

Remember that sites maintained by individuals may or may not be current. The data is worth just what you paid for it! Check any dates on the Web page and have a couple of sites to compare before staking a whole day's scanning on just one site.

Finding frequencies after you get to the track

After you get to the track, you can find vendors offering race-day frequency lists for a few dollars. Are they worth it? If you subscribe to a frequency service and just downloaded the latest data, probably not. Otherwise, because the cost is about the same as one of your favorite race-day beverages and the information is kept up-to-date, I'd recommend that you pick up a list.

Before you go to the race, practice editing frequencies and adding channels on your scanner so that it's easy for you to make adjustments at the track. While you're trying to keep an eye on who's making a move down the back straightaway is not the time to be looking up instructions in your operator's manual.

Using a racing scanner

With so many racing fans wanting to listen in, manufacturers are starting to offer scanners customized for race day. These scanners have many of the same features of their general-purpose cousins, but their operating controls are set up in a way that makes using them at the track much easier. Figure 14-1 shows the keypad of the Uniden SC230 Sportcat. This scanner is similar to that of the popular BC246T scanner I discuss in Chapter 10. (Review Chapter 10 for pointers on basic scanner operating if you're a new scanner user.)

The SC230 Sportcat is designed to be set up so that the channel used on the scanner is the same as the number of the car. For example, store the frequency information for Dale Earnhardt, Jr. at Channel 8 (that's his car number) and set up the label to read *Earnhardt.* When the scanner finds Channel 8 active, it stops and displays both the car number and driver's name.

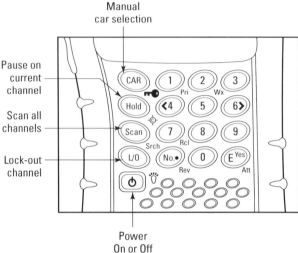

Figure 14-1:
The keypad
of the
Uniden
SC230
Sportcat
scanner is
optimized
for scanning
at the
racetrack.

Manual
car selection

Pause on
current
channel

Scan all
channels

Lock-out
channel

Power
On or Off

When you're at the track, it's a good idea to review your scanner setup before the checkered flag goes down. That's the time to correct any frequency errors, edit driver or team names, and search for new channels not on your lists.

Knowing what frequencies and channels your scanner should cover

To be useful at the racetrack, your scanner should scan from 450 to 470 MHz and from 806 to 956 MHz. It should have at least 100 channels, although a 200-channel scanner is preferable. Although there certainly aren't that many cars on the track (at least not after the first lap, there won't be) you need those extra channels so you can listen to officials, track operations, and broadcasting.

You're better off with too many channels than too few. You will be surprised how fast you can fill up the channels — first with cars, and then with officials and announcers. Next you'll want to have a weather channel or two, and then a local broadcast channel of interest. You can fill up a scanner's channels pretty quickly.

Winging It at Air Shows and Fly-Ins

Air shows bring individual pilots and planes, teams of flyers, and their fans together. *Fly-ins,* like conventions, are events that bring pilots and aviation enthusiasts together, usually without a large public show. You receive a day or two of exciting action — plus the opportunity to get an up-close-and-personal look at some amazing aviation hardware. To really get involved in the spirit of an aviation event, be sure to bring your scanner! Chapter 12 tells you about how to monitor aviation channels, both for planes and ground stations.

To listen in on an air show that involves military planes, your scanner must be able to tune the military aviation bands. Chapter 13's section on military aviation is a must read! Be sure your scanner receives the standard Aircraft band (118 to 137 MHz) and military aviation bands (225 to 389 MHz).

To find air shows, enter the name of local or regional airfields and the search term *air show* or *airshow* into an Internet search engine. There are also Web sites that provide lists of air shows — some shows even have Web pages of their own. Try the aviation directory Thirty Thousand Feet (`www.thirty thousandfeet.com/events.htm`) for starters.

Conducting airfield communications

The first thing you should do is find all the frequencies for the airfield at which the show will be held. The Airnav.com site provides excellent information `www.airnav.com`. Flight control operations are conducted on the usual frequencies, although at a much faster pace, during the *fly-in* and show periods.

If the air show has a Web site, look it over for references to fly-in information or aviation links. These references may give extra information on secondary frequencies that will be used to coordinate airfield traffic. Because of the larger number of planes coming and going, you can expect the Unicom and Multicom frequencies (see Chapter 12) to be busier than usual, as well.

On the ground, security and airfield operations teams are going to be active — probably on the VHF band (148 to 162 MHz) and UHF band (450 to 470 MHz). Use your scanner to search through those frequencies for active channels. Vendors and flight crews will be using FRS frequencies (see Chapter 4), as well.

Listening in on performer communications

The real reason you go to the air shows is to see the astounding performances by skilled pilots as they fly in close formation, do aerial acrobatics, daredevil stunts, and even comedy routines. Along with regular airfield coordination, radios frequencies are set aside for show communications; channels are also set aside so that the team members can talk to each other.

The frequencies for most teams are easily available online by entering the team name of the team into an Internet search engine. Several sites have lists for multiple teams. A comprehensive site with plenty of discussion is hosted by Monitoring Times at `www.monitoringtimes.com/html/mtairshows.html`.

If your scanner has the capability to create and label frequency groups, program each team's frequencies into one group with the team's name. That way you can easily switch frequency sets as the performers change.

Taking Your Radio on the Run and into the Crowd

On any given weekend in a moderately sized town, you can find foot and bicycle races going on, especially in the warmer spring and summer months. Out in the open spaces, road rallies and off-road events are common. Almost all these events use radios to keep things moving along. Because the courses are big, it's almost always impossible to see the whole course at one time; listening to the radio traffic can keep you informed about who's ahead and who's not.

Using your radio at a race

Figure 14-2 shows the elements that are common to most races. Along with the organization that sponsors and runs the race, you find a set of race officials. These people are there to answer the constant stream of questions about procedures and infractions.

Waypoints are stations along the course where race staff keep tabs on the competitors, dispense supplies, and answer questions. In some races, such as rallies, the timing information for each competitor is recorded and passed back to the race committee. Of course, there are first-aid stations. And for the stragglers and those unable to continue there's the *sag wagon,* which comes around to pick them up for a ride home. All these parties must be able to communicate with each other.

Although mobile phones are sometimes used, there is almost always a common channel used for race-wide communications. A spare channel or two may be used for the larger races. The best place to find these frequencies is to search the VHF band (148 to 162 MHz) and UHF band (450 to 470 MHz). If the course is spread out over more than a mile or two, the radios may use a repeater, as described in Chapter 2.

Off-track automobile and motorcycle events often cover a wide area. If repeater coverage of the course is not available, these races may be coordinated by Citizens Band (CB) radio (which I cover in Chapter 5) or amateur radio (which I cover in Chapter 9).

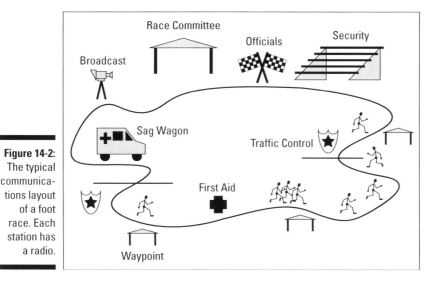

Figure 14-2:
The typical communications layout of a foot race. Each station has a radio.

Law enforcement is also involved in these events, providing security and making sure things run smoothly when the course crosses or uses a public thoroughfare. Law enforcement officials use the regular system or channels for that city or county. Medical personnel also use the designated municipal channels. See Chapter 11 for information about receiving these conversations.

Don't forget the broadcasters! Large races that are televised or broadcast by radio use radio to relay the audio from the announcers, just as described in the auto racing section above.

Using radio technology at a concert or convention

Anywhere a crowd is expected, you'll also find scanning opportunities. Entertainment events often add another element for your pleasure — *wireless microphone systems!* Wireless microphones are low-power radios whose signals can be easily received by a portable scanner over distances of up to several hundred feet. Look for them throughout the VHF band (148 to 225 MHz) as well as in unused UHF TV channels above 470 MHz.

This is a lot of spectrum to search, but the microphones are on continuously when in use, so they're hard to miss. (Frequency assignments vary widely between countries, so check with a microphone dealer or professional sound service in your area.) By listening directly to the microphone's signal, you can listen to one instrument or performer. Explaining everything you wanted to know about wireless microphones (and more), the Shure Company hosts a detailed discussion at www.shure.com/booklets/wireless/wireless_page4.html.

Abide by any restrictions on receiving posted by the promoter or sponsor. Also, you are not permitted to record and distribute what you hear without permission — that's copyright infringement.

Occasionally, the signal fed to a public address or concert sound system is fed to a low-power transmitter for reception by the hard of hearing. You can use a scanner with a set of headphones or earphones to receive this signal.

Getting the Right Accessories

For the events discussed in this chapter, you'll be using your scanner away from home and probably outside. You need the proper accessories. Here's a list of the essentials:

- **Headphones:** Headphones are nonnegotiable if you are bringing a scanner to an auto race. If you have attended a race in the past, I need not say another word (if you didn't wear headphones, you might not hear me anyway). If you've never been, take my word for it: You need to protect your hearing!

  To have a prayer of hearing your scanner's output at a race, you'll need not just headphones, but *full-cup* headphones with soft cuffs that seal around your ears. Open-air foam cushion models let in far too much noise. Away from the racetrack, earphones or lightweight music-player style phones will work fine.

- **A sturdy carrying case:** Outdoor and nighttime use can be tough on electronics of all sorts. Start with a good quality case or carrying pouch for your scanner to protect it from the elements and a clumsy owner! Camera cases and insulated lunch bags are inexpensive possibilities.

- **A dry cloth:** Carry a dry cloth in a sealed plastic bag in case the radio gets wet.

- **Belt clips and neck straps:** These accessories come in very handy when you're lugging a scanner around for a day.

- **A small flashlight:** If your scanner doesn't have a dial light feature, a key-chain solid-state blue or white light allows you to see what you're doing.

✔ **Spare batteries:** Can you imagine showing up to the race or airfield, turning on the scanner, and having it fizzle out on you? You'll need at least one set of spare, fully charged batteries, particularly if you're planning an all-day excursion with lots of listening time. I cover batteries and battery packs obsessively in Chapter 19.

If you use rechargeable batteries (which I highly recommend), fashion battery holders with a set of batteries for your scanner. Think up a way to tell which sets are charged and dead with a label or other method. A quick-charger that can work from a car battery can really save the day, too!

Want to Get Involved?

Scanner manners dictate that you should be a listener and not a player, remember? Still, after some listening, you may find yourself saying, "I could do that!" Know what? You're probably right! You could be the voice on the radio, helping out at a race or at some other public event. Here's how:

✔ Find one of the event organizers and ask if the organization needs volunteers for the next big event. Your offer will be gladly accepted for a future event more often than not.

✔ If you see a notice in the newspaper of an upcoming event, contact the sponsors and ask the same question.

✔ If you're a member of a scanner club, ask around to see whether anyone else participates in public service activities or other radio clubs.

✔ Police and fire departments often have auxiliary groups you can join to help out with traffic control or other minor duties during events ranging in size from local carnivals to state fairs. Scouting groups also assist with public events; you may be able to team up with these organizations to help out.

You'll enjoy working with the other volunteers and competitors and spectators alike will appreciate your public service.

Scanning Tips

In no particular order, here are some good ideas for using your scanner at all kinds of public events:

✔ **Make a cheat sheet:** Copy the pages in your operating manual that cover common functions, such as editing a frequency or performing a frequency search.

✔ **Take advantage of the Priority function:** Use your scanner's Priority function to listen to your favorite channel more frequently.

✔ **Lock out undesired channels:** After the event is underway, use the Lock-out function to skip channels used by competitors that have withdrawn or that you decide you don't want to listen to.

✔ **Practice common operations before the day of the event:** For example, if you're heading to a car race, practice pausing the Scan button and selecting one channel. You'll be doing this a lot during the event, so you may as well get used to it now.

✔ **Bring headphones or ear buds:** At a loud event like a race, you'll need them to hear. At a quiet event, they prevent you from annoying your neighbors.

✔ **Reduce the signal levels:** Have a stubby antenna at the ready or use the scanner's *attenuator* to reduce the overall level of signals coming into the scanner. You'll be very close to the transmitters and really strong signals can bleed through into other channels, causing interference.

Chapter 15

Surfing the Air World: Shortwave Listening

*I*f you're reading through this book chapter by chapter, you have probably noticed that many frequencies below 30 MHz are unaccounted for. Maybe you picked up a radio with several unfamiliar bands marked *LW* or *SW* and wonder what is broadcast at these frequencies.

Welcome to the shortwave (SW) bands between 2 and 30 MHz! Popular outside the U.S. for many years, shortwave broadcasting is enjoying a resurgence of interest here due to inexpensive, high-quality receivers and a heightened interest in world affairs.

Tuning in to SW signals is called *shortwave listening* and the people that do it are *shortwave listeners* or *SWLs*. (These signals are called shortwaves because in the days before TV and FM, they had the shortest wavelengths in common use.) *Longwave (LW)* and *medium wave (MW)* signals are at frequencies below the shortwave stations and so have longer wavelengths.

There is a wealth of signals on the shortwaves from broadcasters, radio amateurs, government and military stations, ships, and planes. This chapter explains how to find shortwave bands on your receiver. Then I go over how you use a shortwave receiver and set up an antenna for it. I discuss how shortwave signals get from one point to another and what affects their travels — and therefore the quality of your reception. I also show you how to use the information in program guides and schedules. After capturing the signal of a station or two, you'll be ready to start collecting confirmations from around the world.

Finding Shortwave Broadcasters

Maybe you've heard of the German radio broadcaster Deutsche Welle or the Dutch radio broadcaster Radio Nederlands? These are two of the best-known SW broadcasters. The U.S. has its own — Voice of America. These powerful stations put out megawatt-sized signals, but similar broadcasts from many smaller countries and private stations fill the airwaves around the clock. You'll find three basic SW broadcast types on the air:

- ✔ **International:** Operated by a national government, these stations promote a country's official voice on the airwaves. These organizations broadcast news, music, and cultural programming aimed at foreign listeners, often in several languages. They also provide programs in the native language of the country for expatriates — a great way to brush up on a language. Deutsche Welle, Radio Nederlands, and Voice of America are examples of international broadcasters.

- ✔ **Domestic:** In a number of countries, particularly large ones, SW broadcasts replace the AM broadcasts we are familiar with in the U.S. Domestic SW programs are usually broadcast entirely in the country's native language and consist of news, weather, sports and other items of local interest. The Russian station Radio Rossii broadcasts in North Asia are for domestic consumption, for example. These stations are found on lower shortwave frequencies and on the LW and MW bands.

- ✔ **Religious:** operated by churches and supporters, these stations broadcast religious programming or cultural programs with a religious theme. You can also listen to worship services via these stations. The famous HCJB from Quito, Ecuador, is probably the best-known and one of the oldest religious broadcasters.

Most stations have Web sites that list operating frequencies and program guides. To find an international station, try entering the name of a country you're interested in, along with the word *radio* into an Internet search engine. For example, type *radio poland* to find broadcasts in Poland.

You can also get printed program guides and directories, such as the references at the end of the chapter. These program guides are usually fairly up-to-date. If your interest is in news or music from a specific country or region, search the directories for national stations and then review the program guides.

You may be asking why not just listen to programs over the Internet? There are several reasons to listen via radio rather than your computer:

- ✔ Although a number of broadcasters use the Internet, most don't, including many of the most interesting stations.

- ✔ Traveling with a computer often isn't practical, and hooking up to a broadband Internet connection without paying a lot of money isn't likely.

On the other hand, a SW radio is easy to pack along to receive news in your native language.

✔ SWLing is *fun!* There is nothing like listening to an interesting music program received from halfway around the globe.

Most SW broadcast signals are organized into several bands sprinkled throughout the SW or HF spectrum between 2 and 30 MHz (see Table 15-1). Each band is referred to by the approximate wavelength of its signals. (1 kilohertz or kHz is ⅟₁,₀₀₀ of one MHz).

Table 15-1	The Shortwave Broadcasting Bands		
Band	*Frequencies (kHz)*	*Band*	*Frequencies (MHz)*
120 meters	2300 to 2495	25 meters	11.500 to 12.160
90 meters	3200 to 3400	22 meters	13.570 to 13.870
75 meters	3900 to 4000	19 meters	15.030 to 15.800
60 meters	4750 to 5060	17 meters	17.480 to 17.900
49 meters	5730 to 6295	16 meters	18.900 to 19.020
41 meters	6890 to 6990 and 7100 to 7600	13 meters	21.450 to 21.750
31 meters	9250 to 9990	11 meters	25.670 to 26.100

Why so many bands? The frequency of the signal has a big effect on its ability to travel long distances at different times of the day or in various seasons. By using different bands based on climate and other factors, SW broadcasters can get their signals to the right area at almost any time.

Listening to amateur radio on SW bands

Located on SW bands mixed in between (and sometimes shared with) those of the broadcasters, you can find amateur or ham signals night and day from all corners of the earth. (Amateur radio is discussed in Chapter 9 and in my book, *Ham Radio For Dummies*.) You can tune in personal conversations and listen as hams *chase DX* (DX means *distant station*, which are sought-after contacts); you can also monitor message relaying groups called *nets* and participate in competitive contests as a receiving station.

Bringing bits to shortwave: Digital Radio Mondiale

As you browse the directories and Web sites of SW broadcasters, you may encounter DRM or *Digital Radio Mondiale*, a new standard for propagating SW signals. Many of the larger broadcast stations make DRM transmissions and more are joining them every year. DRM is really the name of the international consortium created to bring digital technology to AM and SW broadcasting. Fading and interference can often be an impediment to clear reception, particularly at long distances. DRM technology goes a long way to overcoming these problems, bringing an FM-like quality to AM signals. Because DRM signals are digital, they can also carry digital data, providing additional information about the program or the broadcast station.

DRM should not be confused with Digital Audio Broadcasting (DAB), which is an enhancement of FM broadcasts, or with digital TV signals, such as DirecTV. Although DAB and DRM signals are delivered as streams of digital data, the two formats are much different. If you'd like to know more about DRM, visit the consortium's Web site (www.drm.org). You can also find more information from Wikipedia (www.wikipedia. org). You can also visit the FAQ pages of the major international broadcasters. For example, Radio Nederlands offers an article called "A Listener's Guide to Digital AM (DRM)." Visit the Web site (www2.rnw.nl/rnw/en) and type **listeners' guide DRM** into the search box.

Most SW amateur signals are a variation of AM called *single sideband* or *SSB*. (SSB is discussed in Chapter 5.) You need a radio with sideband capability. On frequencies below 10 MHz, SSB transmissions use lower sideband or LSB and elsewhere upper sideband or USB. Hams also use Morse code (abbreviated CW), radioteletype (RTTY) and other digital modes discussed in the following section. Table 15-2 shows a list of frequencies used by amateurs in the shortwave bands. Many of these bands are adjacent to SW broadcast bands.

Table 15-2	The Shortwave Amateur Bands		
Band	*Frequencies (kHz)*	*Band*	*Frequencies (MHz)*
160 meters	1800 to 2000	20 meters	14.000 to 14.350
80 and 75 meters	3500 to 4000	17 meters	18.068 to 18.168
60 meters	5330.5, 5346.5, 5366.5, 5371.5, 5403.5 (USB)	15 meters	21.000 to 21.450
40 meters	7000 to 7300	12 meters	24.890 to 24.990
30 meters	10100 to 10150	10 meters	28.000 to 29.700

Picking up MW and LW broadcasts

U.S. AM broadcasters from 550 and 1700 kHz occupy a band known to SWLs as medium wave (MW). There is a similar broadcast band outside the U.S. SWLs seek out distant MW stations just as they look for higher-frequency SW broadcasts. You can do some *broadcast DXing* (tuning your radio to receive distant AM stations) with an ordinary portable radio by attaching some antenna wire to it or by driving out to a quiet location and using your car radio. The best time to listen for long-distance MW reception is at night, when the local stations go off the air or reduce power so that the *clear channel* stations can be heard for hundreds of miles. Even lower in frequency are the longwave (LW) stations that broadcast in Europe, Asia, and Africa. They are very difficult to receive in North America, because these frequencies are used for beacon stations.

Within each amateur band, the voice, code, and digital signals are organized by a *band plan*. The complete list of these band plans is available for U.S. stations on the American Radio Relay League's (ARRL) Web site at www.arrl. org/FandES/field/regulations/bandplan.html. Outside the U.S., the band plans are published by the International Amateur Radio Union (IARU) at www.iaru.org/bandplans.html.

Monitoring commercial, government, and military broadcasts

If you have a *general-coverage* receiver that can tune between the broadcast and amateur bands, you will find a lot of interesting traffic to monitor. Ships and planes use HF (the nonbroadcast term for shortwave) signals when they are too far away from land to take advantage of short-range land VHF stations. Land stations are called *fixed* or *mobile* stations and use HF channels for regional and long-distance communications.

Chapter 13 focuses on receiving short-range military signals in the VHF and UHF spectrum. Military transmissions are found in the HF spectrum, as well, because the signals provide good regional and long-distance coverage without relying on any intermediate repeater or relay stations. You can hear everything from military bases to planes, ships, vehicles, and even individual soldiers talking to each other.

The International Telecommunications Union (ITU) ensures that each group of users gets access to enough of the spectrum (these parcels of spectrum are called *allocations*), sometimes sharing it with other users. A table of allocations is available at radioscanning.wox.org/Scanner/miscfreqs/HF/us_hf_frequency_allocations.htm. Conducting a search for the term *hf frequency allocations* at an Internet search engine also turns up quite a few reference sources.

The Monitoring Times provides a list that it calls the "Hot 1,000 Frequency List." Check it out at www.monitoringtimes.com/html/mttopHF.html. The list includes active maritime, aviation, government, military, and commercial channels. If you find yourself frequently visiting these frequencies, you might want to invest in a receiver that has the capability to act as a scanner. Happy hunting!

Using shortwaves to receive data and miscellaneous signals

You can listen to several other types of communications in addition to voice conversations, of course. With your shortwave receiver, you can also find stations sending data, pictures, and other types of information. If you're up to it and have the right equipment, you can decode and read this information!

Radioteletype and digital signals

The first time you hear what it sounds like when information is transmitted as digital streams of data, it may sound strange. If you've ever sent a fax or listened to a modem hookup, you might get a sense of what these transmissions sound like to human ears.

Here's a list of complex types of digital signals:

- **CLOVER:** Data transmissions using a signal with four different frequencies
- **Fax:** Facsimile, the same transmissions that your fax machine sends by phone
- **MFSK (Multi-Frequency Shift Keying):** Text sent as a closely spaced group of varying tones
- **NAVTEX (Navigation Text):** Weather and other information sent to ships
- **RTTY (Radioteletype):** The oldest method of sending character messages

On the SW bands, data is transmitted as tones. Radioteletype (RTTY) signals, for example, are composed of two tones that are transmitted in an alternating pattern, changing several times per second.

You must have the right software to recover or *demodulate* the data from each of these types of signals. Signal demodulation software is listed at both `ac6v.com/software.htm#DIGITAL` and `www.dxzone.com/catalog/Software/DSP`. If you decide to get serious about monitoring digital signals, Skysweeper (`www.skysweep.com/index.html`) is a top-of-the-line program that receives and demodulates data in many different modes.

Listening to beacons, WWV, and standard stations

You may run across *beacon* stations that transmit a continuous tone with an occasional burst of Morse code or data. Other stations, such as the U.S. standard station (call letters WWV) at 2.5, 5, 10, 15 and 20 MHz, transmit a regular, one-second tick or data burst — these are time and frequency standard stations.

Beacons are used for navigation or to assess how well signals are propagating from one place to another. By measuring the signal of two or more beacon stations, ships and planes can determine their precise location. The relative strength of beacon signals tells the receiving operator how good (or poor) conditions are for signals transmitted from the beacon's location.

Standard stations are handy for setting clocks and keeping time. For example, the atomic clocks and wristwatches listen to a very low-frequency standard station to keep time. The regular data bursts can be received and interpreted by a computer to adjust its clock. The very precise frequency of the standard stations is also used to calibrate receivers and other equipment.

Choosing and Using SW Radios and Antennas

You can find a wide variety of receivers for your adventures in SWLing. Some radios are very simple — not much different than the ubiquitous AM/FM models — while others are full-blown commercial-grade receivers with a front panel full of knobs and dials.

The more capable receivers can be quite confusing to a beginner, so if you're just starting out, choose one of the midrange portable models that receives shortwave broadcasts, such as the Grundig YB400PE.

Selecting a radio

SW receivers come in many different styles, from handheld models (some even fit into a shirt pocket!) all the way to desktop signal crunchers. If you're just interested in casual listening, a simple portable unit will do nicely. When you're fully sucked in to the shortwave world, you may find yourself becoming interested in pulling in weak signals on difficult bands. That's the time to move up to a *general-coverage* or *communications-grade* receiver.

Here are specifications you should know:

- ✔ **SW broadcast-only radios:** These radios don't tune outside the SW broadcast bands. As long as you're only interested in broadcast signals, these radios work just fine.

If you plan on SWLing regularly, make sure that *any* radio you purchase can handle Digital Radio Mondiale signals. Because DRM signals, (which I describe in the sidebar, "Bringing bits to shortwave: Digital Radio Mondiale," earlier in this chapter) are being added by many broadcasters, getting a radio with this capability will prevent early obsolescence.

- ✔ **General-coverage receivers:** These receivers tune from below 200 kHz to 30 MHz and receive AM, USB, LSB, CW (continuous wave or Morse code), and FM signals. General-coverage receivers offer extensive memory capacity and advanced tuning features, such as *sweeping* (tuning through a range of frequencies) and *scanning* (listening to a sequence of preprogrammed frequencies). See Chapter 10 for more information about sweeping and scanning.

If you think you'll eventually want to go beyond broadcast reception, it's well worth the extra expense for a good general-coverage receiver. The circuit design is better and the receiver will perform better with regards to noise and interference rejection.

- ✔ **Communications-grade receivers:** These receivers have super performance specifications and advanced computer control capability. They tune through the VHF and UHF bands, as well, and often perform double duty as high-performance scanners.

Universal Radio publishes a comparison chart that provides a good look at the capabilities of both general-coverage and communications-grade receivers — www.universal-radio.com/catalog/widerxvr/chartw.html.

I recommend visiting the IBS (International Broadcasting Services) Web site. IBS publishes the best-selling shortwave radio guide in the universe (as far as we know). This publication, *Passport to World Band Radio* is published annually (visit www.passband.com) and is available for download (of course, you have to pay IBS money for this guide, but the information is well worth the price).

Introducing common SW radio controls

Imagine that your ordinary AM/FM portable radio is able to receive more bands, can store a whole directory of station frequencies, and scan the band for stations. Those are the features of the popular portable SW receivers, such as the one in Figure 15-1. When you're comfortable with the new frequencies and tuning options, you'll feel right at home. Here's the short list of controls and functions:

- **Mode:** Select AM, FM, or SSB directly or cycle through a list of supported signal types. SSB may include both USB and LSB settings.

- **Frequency Entry:** If you know what frequency you want to listen to, enter the frequency directly instead of tuning.

- **Lock and Reset:** Lock the keypad to prevent accidental changes. Reset clears the radio's list of stored frequencies.

- **Memory:** When a station is tuned in, you can store the frequency in a directory for recall later.

- **Tuning:** Change frequencies either in steps or by automatically sweeping through a band. A fine-tune control (sometimes called a *clarifier*) is offered on some radios.

- **S-Meter:** Signal strength is measured in *S-units*. A meter with a moving needle or an LCD bar graph are the most common strength indicators.

The Grundig YB400PE radio has a control on its side that enables you to tune between USB and LSB. It also has jacks for external antennas, headphones, and a power supply.

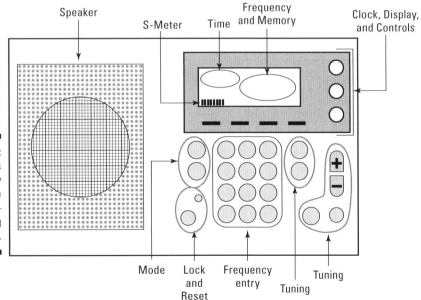

Figure 15-1:
The controls of a quality portable receiver — the Grundig YB400PE.

Getting a shortwave antenna

The most common SW antennas are *dipoles* (lengths of wire with the feedline attached at or near the middle) and *random-wires* (the feedline is attached at or near the end). In general, the higher the antenna, the longer the range it will have on both transmit and receive. These antennas aren't fussy about where they are installed. Other types of antennas include

- **Vertical:** This type of antenna is usually mounted to a metal structure that serves as a ground. It acts similarly to half of a dipole turned on its side. Verticals receive signals equally well in all directions.

- **Active antennas:** These antennas consist of a small loop or a short vertical element *(whip)* with an RF amplifier built in to compensate for the short antenna. Active antennas will do if you can't put up a full-sized antenna, but they're usually a compromise in performance.

- **Trapped:** This antenna includes tuned circuits (the *traps*) to adjust the effective length of the antenna automatically.

- **Loops:** Primarily used at frequencies below 5 MHz where a full-sized antenna would be too large. A well-designed loop can work much better than a wire antenna at these low frequencies.

Radio dealers usually have a good collection of suitable antennas for sale. For more information on which antennas are best for SW reception, visit www. ac6v.com/swl.htm#ANT for a good collection of articles and Web pages to try. The ARRL's Technical Information Service page (www.arrl.org/tis/ tismenu.html) is an excellent source of information, as well.

A decent, external antenna makes a big difference in your receiver's performance. If you already have an antenna but you're not happy with its performance, simply upgrading to an improved antenna delivers a lot of bang for the buck. Be sure your radio has an external antenna connection and use it if you can.

Building your own antenna

If you like to build things, wire antennas make an excellent project. The materials are inexpensive and performance is good. If you'd like to try building an antenna, the Do It Yourself Web site www.diymedia.net/links/lschem listen.htm has several simple antenna projects. Numerous antenna articles and links can be found at www.dxzone.com/catalog/Antennas/Shortwave and the book *Easy-Up Antennas for Radio Listeners and Hams* by Ed Noll (Radio-Canada International) is a good way to get going.

You can install antennas inside an attic or under a roofline if you're not permitted to have outdoor antennas. Keep antennas clear of large metal structures that can degrade reception. If you're an apartment dweller, you can hang a temporary wire outside, reeling it back in when your receiving session is done.

Putting up an antenna near power lines of any sort can be dangerous. Never erect an antenna over or under power lines. Keep an antenna far enough away from power lines that it can't blow or fall onto them.

Finding SW equipment vendors

Where do you get this stuff? Here is a list of some distributors of radios, antennas, and general radio supplies:

- ✔ Universal Radio (www.universal-radio.com)
- ✔ Grove Enterprises (www.grove-ent.com)
- ✔ Ham Radio Outlet (www.hamradio.com)
- ✔ Amateur Electronic Supply (www.aesham.com)
- ✔ RadioShack (www.radioshack.com)

These vendors sell premade antennas and antenna materials:

- ✔ Alpha Delta Communications (www.alphadeltacom.com)
- ✔ Radio Works (www.radioworks.com)
- ✔ Radioware (www.radio-ware.com)
- ✔ The Wireman (www.thewireman.com)
- ✔ Davis RF Company (www.davisrf.com)

Shortwave Signal Propagation

One of the unique aspects of the shortwaves is their dynamic propagation from point to point. Signals can travel in one of two ways — as *ground waves* or as *sky waves*. Strangely enough, ground waves travel along the surface of the earth and sky waves are reflected from the sky! Sky-wave signals are also referred to as *skip*, as in, "I'm hearing skip from South America today!"

The AM broadcast signals you receive through the day get to you as ground waves. At the lowest SW frequencies, ground waves can travel as far as 100 miles before finally being absorbed by the earth. (MW and LW signals can go even farther.) As the frequency increases, so does absorption by the earth until, at 30 MHz, the ground wave travels only a few miles.

Sky waves are a bit more complex, depending on the characteristics of the *ionosphere*, thin layers of the atmosphere high above the ground. The ionosphere consists of three layers; D (lowest), E, and F (highest), as shown in Figure 15-2. (The F-layer is really *two* layers, F1 and F2, that act slightly different, but can be considered a single layer for our purposes.)

The ionosphere is created by solar *ultraviolet* radiation, which knocks electrons off of atoms. The result is a weakly conductive layer of gas. The electrically charged particles in the ionosphere can bend passing radio waves. The higher the frequency of the wave, the less likely it is to bend. (Similarly, water bends light waves of different colors by different amounts.)

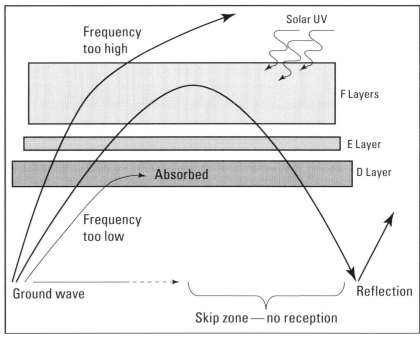

Figure 15-2: The ionosphere reflects SW signals, allowing reception of sky-wave signals at locations too far away for ground-wave propagation.

Hops, skips, and jumps: Understanding how the ionosphere affects propagation

Here's what happens to radio waves in each layer of the ionosphere:

- ✔ **D layer:** Because it is close to the ground, the D layer has so many charged particles that it acts like a radio sponge, soaking up the signals passing through. The lower the frequency of the signal, the more the D layer absorbs, putting a lower limit on the frequency of signals traveling as sky waves. Frequencies above 5 MHz generally penetrate the D layer and continue skyward.

- ✔ **E layer:** Signals above 5 MHz survive their trip through the D layer only to encounter the E layer. The weak E layer has little effect on passing signals, only absorbing and bending them a little bit. Occasionally, an intense patch of E layer will form that can reflect signals like the F layer — propagation from these reflections is called *sporadic-E*.

- ✔ **F layer:** Most of the bending takes place up in the F layer. If the frequency is too high, not enough bending takes place and the wave escapes into space, never to be heard again. At lower frequencies, the bending can be enough to turn the radio wave back toward the earth, where it can be received again as a sky-wave signal.

The ionosphere's ability to bend a wave also depends on the angle at which the wave enters the ionosphere — low-angle waves are easier to bend. The trip up to the ionosphere and back again is called a *hop*.

Because the earth also reflects waves, if the frequency and angle are right, multiple hops can allow a signal to travel thousands of miles, even all the way around the earth! This is *skip propagation*. When signals propagate by sky wave, the band is said to be *open*. If the first hop is longer than the maximum groundwave distance, there is a donut-shaped area centered on the transmitting station in which no signal at all can be heard. This is the *skip zone*.

Signals received as a result of one or more hops are considered *DX*, short for *distant station*. Receiving DX signals is one of the most popular of all SW activities!

Because the ionosphere is created by sunshine, reflected signals travel farther during the day and weaken or disappear at night. This is especially true of signals at the high end of the SW bands — there isn't enough F layer left to reflect them after night falls. Conversely, when the D layer disappears, low-frequency signals are no longer absorbed. As a result, low-frequency signals begin to travel as sky waves at night. This is why AM broadcasts and other

MW and LW signals are received best at night. Here's a day-and-night look at signal propagation:

- ✔ **Daytime:** High-frequency signals can travel farther as sky waves because they are reflected off the F layer of the ionosphere; low-frequency signals are absorbed by the D layer of the ionosphere and therefore don't travel as far.

- ✔ **Nighttime:** High-frequency signals don't travel far because the diminished F layer of the ionosphere does not reflect them; the ionosphere no longer absorbs low-frequency signals because the D layer is diminished, so they travel farther as sky waves.

Knowing where the world is in day and night helps you understand short-wave propagation. There's a great map of the *terminator* (the line between day and night) at www.worldtime.com.

Table 15-3 shows how signals of different frequencies propagate at different times of day.

Table 15-3	Daytime/Nighttime Shortwave Band Behavior	
Band	*Daytime*	*Nighttime*
Below 5 MHz	Local and regional out to 100 to 200 miles.	Local to long distance with DX near sunset or sunrise.
5 to 10 MHz	Local and regional out to 300 to 400 miles.	Short-range (20 or 30 miles) xand medium distances (150 miles) to worldwide.
10 to 20 MHz	Regional to long distance. Bands will open at or near sunrise and close at night.	Often open in the West in the evening and may be open 24 hours a day during the solar maximum.
20 to 30 MHz	Primarily long distance, opening East after sunrise and West in the afternoon.	Bands are mostly closed at solar minimum, but open every day at solar max.

If you really want to know the details of shortwave propagation, a detailed tutorial "Introduction to HF Radio Propagation" is available at no charge from the Australian Space Weather Agency at www.ips.gov.au. (Click the Educational link; then look for an Other Topics heading. Click the Radio Communications link under that heading.) The ARRL Technical Information Service also presents a number of good links and articles at www.arrl.org/tis/info/propagation.html.

Understanding other atmospheric conditions that affect propagation

Radio propagation is affected, not just by the rising and setting sun and other daily variations (see the previous section), but also seasonably by the tilt of the earth toward the sun as it moves through its yearly orbit. The changing solar illumination also changes the maximum and minimum frequency at which signals can propagate from point to point at different times of the year because the ionosphere is formed differently at different seasons. SW stations change their transmitting frequencies and schedules to compensate for this effect.

Along with daily and seasonal variations, the amount of solar radiation reaching the ionosphere also has a big effect on SW signals. The amount of solar UV varies dramatically with the 11-year *sunspot cycle*. The more sunspots on the sun, the more solar ultraviolet light that is radiated. Higher sunspot years (the last peak or *solar maximum* was in 2002) are great for signals above 15 MHz — they can often be received 24 hours a day or deep into the night because the extra UV light keeps the F layer energized. Those years are not so good for low-frequency signals because the D layer is also energized and has no problem soaking up low-frequency signals. Savvy SW stations also know to account for these variations in propagation by changing their frequencies appropriately.

Introducing World Time

Because SW signals are often received thousands of miles away, keeping time zones straight can be very difficult to manage. That is why the international radio community has standardized on using a single 24-hour clock, called *World Time.* The clock keeps the same time as London, also referred to as *Universal Coordinated Time* (*UTC* is a French acronym) or previously, *Greenwich Mean Time* (*GMT*) abbreviated with a Z. All program guides (see "Using a Program Guide," later in this chapter) use World Time.

To find out how to convert your local time to World Time, use the Web site wwp.greenwichmeantime.com. If you live in North America, click your country or region (the list appears under the Time Zones heading on the left side of the screen; then click the state to find out how far behind (or ahead) of Big Ben you are. Time Zones are shown as the difference in hours from GMT. For example, Pacific Standard Time is eight hours behind GMT, or GMT-8. If you live outside North America, click the Time Zone link first and then click the appropriate maps.

Using a Program Guide

Figure 15-3 shows a sample of a program guide, modeled on the *Passport to World Band Radio* listings (I discuss this guide in "Selecting a radio," earlier in this chapter, and in the sidebar, "Starting your shortwave library"). Directories may be organized in several ways: by frequency, time, or place of origin. In the figure, a listing of English language programs from the Turkish government station Voice of Turkey is shown. Organized alphabetically, the listing for Radio Turkey is preceded by the programs from Radio Thailand and followed by Radio Ukraine.

Each program is broadcast at different frequencies depending on what part of the world it's being beamed to. These frequencies are chosen to maximize the received signal in the target area at that time of day. Seasonal changes are also noted with the winter and summer schedule symbols.

Enter the frequency into an Internet search engine, for example *11625 kHz*, and you can get a list of Web sites of stations that use that frequency!

Frequency and target area

Broadcast station

Figure 15-3: Program guides show SW broadcasts in World Time by country and station. Frequencies vary by season.

Voice of Turkey
0300-0350 S7370 (Mideast),S11665/9650 (Europe & N America)
0400-0450 W6020 (Europe & N America),W7420 (Mideast)
1230-1325 S17595 (S Asia, SE Asia & Australasia, S17830 (Europe)
1330-1425 W15145 (Europe),W5195 (S Asia, SE Asia & Australasia)
1830-1920 S9785 (Europe)
1930-2020 W5980 (W Europe)
2130-2220 9525 (S Asia, SE Asia & Australasia)
2200-2250 S9830 & S12000 (W Europe & E North Am)
2300-2350 W6020 (W Europe),W9655 (W Europe & E North Am)

World time (GMT or UTC)

W Winter schedule
S Summer schedule

Starting your shortwave library

If I've convinced you to give shortwave listening a try in earnest, having a copy of one of these fine books is a necessity:

✔ *Shortwave Listening Guidebook* by Harry Helms (High Text Publications)

✔ *Passport to World Band Radio,* yearly directory (International Broadcast Services, Ltd.)

✔ *World Radio and TV Handbook,* yearly directory (WRTH Publications, Ltd.)

Both the directories also contain listings organized by frequency to enable you to identify a broadcast if you happen across it while scanning the airwaves. You'll find this to be a very useful feature. Find the frequency in your program guide and the stations that are known to transmit at the frequency are listed along with the time of day of their programs. With a little head scratching, you can figure out which station you're hearing.

Confirming Your Reception

Finally! You've obtained a radio, strung an antenna, anxiously tuned around to find a strong signal, and enjoyed your first SW broadcast program! Well done! Now's the time to send a reception report and start your *QSL* collection.

QSL is an old radio abbreviation meaning "received and understood." In modern times, though, a QSL is a postcard or letter that SWLs exchange with stations to acknowledge that the SWL has received or *logged* the station's signals. Figure 15-4 shows a SWL QSL that I received as part of my amateur radio operations in 1997. F11FIK is the call sign of Mssr. Claude Renard, a French SWL who heard me contact a French amateur station, F6ELE. Claude has designed a very attractive card with all the necessary information. Of course, it is standard procedure to send a return QSL or letter confirming the confirmation.

Your QSL is your radio business card and you can send one for every station you hear or just for the most interesting transmissions you pick up. With the excellent, reasonably priced home graphics programs and printers available you can design your own QSL — and many SWLs do. You can also use commercial designs, such as those from `cheapqsls.com` or `www.rusprint.com`. Collecting these cards is very popular and your collection can quickly grow into the hundreds of colorful QSL trophies.

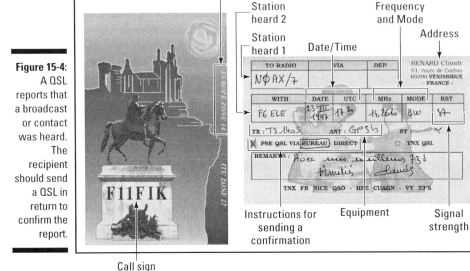

Figure 15-4: A QSL reports that a broadcast or contact was heard. The recipient should send a QSL in return to confirm the report.

Including the right information in your QSL

Your QSL should include the following information:

- The call sign of the station to which you are sending the QSL. If you happened across a contact between two stations, include the call signs of both stations such as, "F11FIK with F6ELE."

- Your call sign, which is usually printed on the card.

- The frequency at which you heard the transmission.

- The time and date of the contact (in World Time always).

- The type of signal (Morse code, SSB, and so on).

- Comments about signal strength or conditions are optional.

If your report is accurate, you should receive a QSL in return confirming your report. If you haven't created a standard card with all the necessary information, a letter will do just fine. To speed up the process and encourage a response, include a self-addressed, stamped envelope with sufficient postage to receive a return QSL.

Getting station addresses

Addresses for SW broadcast stations are available on their Web sites or in program directories. Sometimes, hams exchange mailing addresses over the air, but more often, you'll find their addresses at `qrz.com`, a Web site that accesses government license databases. Simply enter the call sign into the Get Callsign box to view their license information.

SWL Web References

Along with the printed material that I recommend in the sidebar, "Starting your shortwave library," you should know about the large number of Web sites devoted to SW and LW, and MF receiving. The following three sites provide a great deal of information, as well as numerous links to other related and specialized sites. Happy surfing:

- ✔ **DX Zone (`www.dxzone.com`):** A comprehensive list of links related to SWL, ham radio, and CB radio. Click the <u>Shortwave Listening</u> link to focus on SWL activities. A comprehensive list of SWL clubs and organizations is available at `www.dxzone.com/catalog/SWL_BCL/Clubs_and_Organizations`.

- ✔ **AC6V's Amateur Radio and DX Reference Guide (`ac6v.com/swl.htm`):** Another comprehensive site. Click the <u>SWL/SCANNER/BC</u> link to visit a page that focuses on shortwave action.

- ✔ **Hard-Core-DX (`www.hard-core-dx.com`):** A very detailed and thorough site devoted to SWLs with the latest SW broadcast news. It provides an e-mail list for listeners.

Part IV
Getting Technical with Your Radio

The 5th Wave   By Rich Tennant

"We're not sure what it is. Rob cobbled it together from spare parts in his garage. But you wouldn't believe the range this baby gets."

In This Part . . .

*W*hat radio enthusiast doesn't have a need for a little hands-on know how? This part is all about the nuts and bolts of keeping your radio and antennas in top shape. The toolbox is the first stop, and I introduce you to the tools and gadgets that are mighty handy to radio gurus. After that, I present a bit of the electrical savvy that makes radios run and introduce you to some of the parts and pieces that make up radios and radio stations. Finally, I show you how to install your radio wherever you may be — at home or in a vehicle.

You'll definitely get a charge out of the chapter on batteries! A portable radio relies on these power packs to work and you'll want to know which batteries are the best ones for the job. A computer and a radio work together in powerful ways, so I explore the possibilities of hooking up your computer and software to interface with your radio. Finally, I give you some awesome troubleshooting advice so that when Mr. Murphy comes knocking (law books in tow), you can shoo him away.

Chapter 16

Building Your Radio Toolbox

*I*f your radio is pretty basic, you may have purchased everything you need to operate it. Nowadays, many gadgets are simply premade components, so it's quite possible that you'll never need a single tool or part! Radios are quite reliable and well made these days. Your only maintenance may be an occasional changing of the batteries and a cleaning from time to time. If that's what your radio needs to stay happy, good for you!

Many radio users eventually decide to install a radio in their car, put up an outside antenna, or even poke around in the radio itself. It's a big money saver to make your own cables, for example! Knowing your way around a toolbox can pay big dividends if you ever find yourself away from home and need to install a new antenna or power cable. Being able to do your own maintenance and installation doesn't require an engineering degree, but you do have to have the proper tools and parts. As an added bonus, when you realize that you have the right tools and parts, you'll be fixing more than just radios around the house.

Acquiring the Right Tools

You may have some of the tools already, but working on electronics requires mostly different tools than, say, woodworking or engine repair. If you're just getting started, you'll probably have to augment your existing toolset.

Absolutely required tools

Maintenance involves taking care of all your equipment, as well as being able to fabricate any necessary cables or fixtures to put it all together. A good set of suitable tools is shown in Figure 16-1. Having wire tools on hand enables you to do almost any electronics maintenance task. Every toolbox should contain the following items:

✔ **Head-mounted magnifier:** Electronic components are getting smaller by the hour, so do your eyes a favor. Head-mounted magnifiers are often available at craft or fabric stores. There are also clamp-mounted swing-arm magnifier/light combinations. Plus, a head-mounted magnifier is guaranteed to give you that techie-geek appearance you've struggled your whole life to perfect.

✔ **Soldering iron and gun:** Soldering irons are also known as *pencils* or *pens*. Delicate connecters and *printed circuit boards* (PCBs) require you to use a soldering iron that works at a low temperature and has a fine-point tip.

Look for a soldering iron that is rated around 30 watts. Heavier soldering jobs (soldering wires heavier than 18 gauge, for example, or soldering some types of connecters), take more heat. For these you'll need a *soldering gun* that should have a power rating of 200 to 250 watts.

✔ **Terminal and connecter crimpers:** *Don't* use regular pliers on crimp terminals. You might pull the connection out or work it loose; then you'll spend hours chasing down the loose connections. Find out how to install crimp terminals and connecters and do it right the first time with the right tools.

✔ **Volt ohm meter (VOM):** If you can afford it, get a VOM with diode and transistor checking, a continuity tester, and a capacitor and inductance checker. Some models also include a frequency counter, which can be handy.

✔ **Wire cutters:** Use a heavy-duty pair of wire cutters to handle big wires and cable and a very sharp pair of diagonal cutters or *dikes* with pointed ends to handle small jobs.

Not absolutely required, but certainly handy, tools

Of course, the tools I mention in "Absolutely required tools" are just the beginning. Other tools that are handy include the following:

✔ Allen wrench or Hex Key wrench set

✔ Awl and tee-handled reamer for aligning, making, and enlarging holes

✔ Coaxial (often shortened to coax, pronounced *coh-axe*) cable stripper and crimper for installing connecters on coaxial cable

- ✔ Jeweler's screwdriver set

- ✔ Miniature locking pliers

- ✔ Modular connecter crimper and stripper to prepare flat phone-style cables and install the plastic modular connecters found on microphones and telephones

Finding bargain toolsets

Several companies sell sets of tools that are a good bargain. Conduct an online search and use keywords such as *computer toolset, computer toolkit, electronics toolset,* or *electronics toolkit.* Some online vendors list toolsets and toolkits as *tool sets* or *tool kits* (two words), so try your search both ways.

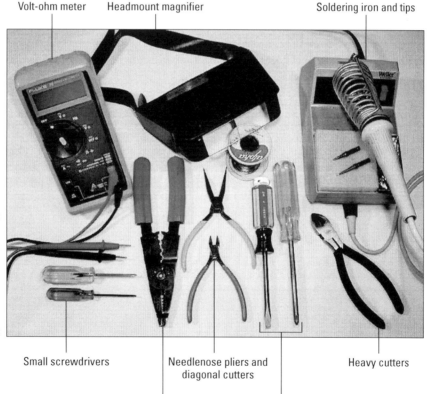

Volt-ohm meter   Headmount magnifier   Soldering iron and tips

Figure 16-1:
A set
of tools
used for
routine
radio main-
tenance.

Small screwdrivers   Needlenose pliers and   Heavy cutters
                                diagonal cutters

Terminal crimper   Large screwdrivers

Home repair and craft kits don't usually have the appropriate tools.

Here are some retailers that sell complete toolkits that will come in handy for your radio:

- **Belkin** (www.belkin.com): Offers a 55-piece toolset with most of what you need.

- **Jameco Electronics** (www.jameco.com): Offers several suitable sets, such as the 20-piece model 170966 that is an excellent starter set.

- **MCM Electronics** (www.mcminone.com): Features several good Duratool kits, as well as its own house brand.

- **RadioShack** (www.radioshack.com): Carries the Kronus line of tools, which includes a number of useful assortments.

Cleaning tools you must have

Cleaning your equipment is an important part of its maintenance. You need the following items:

- **Metal bristle brushes:** Light-duty steel and brass brushes are great for cleaning up oxide and corrosion. Brass brushes don't scratch metal connecters, but they can damage plastic knobs or displays.

- **Soft bristle brushes:** Old paintbrushes (small ones) and toothbrushes are great cleaning tools. I also keep a round brush for getting inside tubes and holes.

   Don't forget to clean off your cleaning brushes! A brush that's dirty isn't really very useful. Get rid of corrosion and grease after every cleaning job.

- **Solvents and sprays:** I use lighter fluid (not butane), isopropyl alcohol, contact cleaner, and compressed air. Lighter fluid cleans panels and cabinets gently and quickly, as well as removing old adhesive and tape. Always test a solvent on a hidden part of a plastic piece before applying a larger quantity.

Getting a toolbox

After you have a set of tools, keep them together in a toolbox and not sprinkled all around the house or shop. I guarantee that repairs are more enjoyable and efficient when you keep the tools together. Figure 16-2 shows my

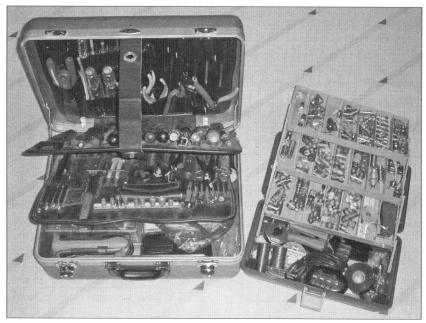

Figure 16-2:
Keep your
tools
organized.

own toolbox. I splurged on a nice toolbox from Jensen Tools (`www.jensen tools.com`) and I have never regretted the decision. The box has containers for each tool. Next to the toolbox is an ordinary plastic fishing tackle box full of RF connecters, adapters, and some related hardware. Using a tackle box is a wonderful option for managing and organizing all your electronic components, tools, and spare parts.

Keeping your stuff in portable kits makes working outside, in your car or boat, or at a friend's house very easy and convenient.

Stocking Stuff

You'll eventually need spare parts, so I suggest that you begin collecting the extras you're most likely to need. The following sections give you the lay of the land.

Stocking extra adapters and connecters

First and foremost, obtain a spare for all of your equipment's connecters. Look over each piece of gear and note what type of connecter is required. When you're done, head down to the local electronics emporium and pick up one or two of each type.

To make up coaxial cables, you need to have a few RF connecters on hand. Be sure to choose the types of RF connecters your radio uses.

You might need an adapter when you don't have just the right cable or if you have a new accessory that uses a type of connecter you don't have on hand. Table 16-1 shows the most common adapter types. Figures 16-3 and 16-4 show what the various adapters look like. You don't have to get them all at once, but this is a good list to have when shopping or to consult when you need an extra part to make up a minimum order.

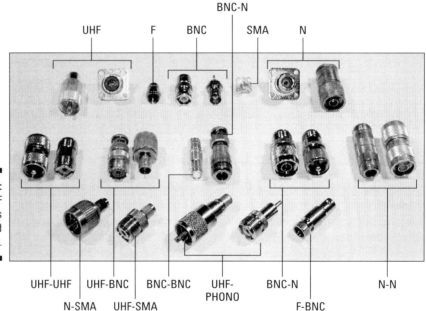

Figure 16-3:
Common RF connecters and adapters.

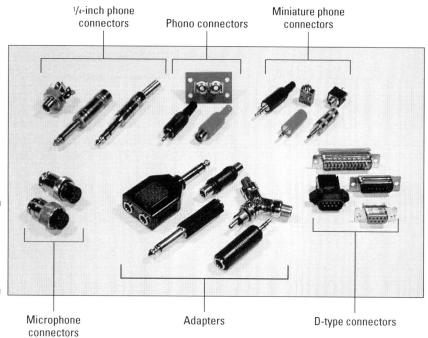

Figure 16-4:
Common
audio and
data
connecters.

¼-inch phone
connectors

Phono connectors

Miniature phone
connectors

Microphone
connectors

Adapters

D-type connectors

Table 16-1	Common Radio Adapters
Adapter Use	*Common Types*
Audio	Mono-to-Stereo phone plug (¼ inch and ⅛ inch), ¼-inch to ⅛-inch phone plug, right-angle phone plug, phone plug to RCA (phono) jack and RCA to phone, RCA double-female for splices, phone plug and RCA plug stereo splitters.
Data	9-pin to 25-pin D-type, DIN-to-D cables, null modem cables and adapters, 9-pin and 25-pin double male/female (gender benders), USB-PS/2 adapter.
RF	Double-female (barrel) adapters for each type of connecter used by your equipment and RF cables. I like to have an adapter between plugs and jacks for each of the styles of connecter. If your radio has a manufacturer-specific connecter, get an adapter to one of the standard types.

Stocking other odds and ends

Along with connecters and adapters (see "Stocking extra adapters and connecters"), there are some common consumable parts and material that you should have on hand. For example, keep the following items on your list:

- ✔ **Cable ties:** These plastic, locking straps are incredibly useful. I've used them to temporarily replace U-bolts and hose clamps. Be sure to get some black, ultraviolet (UV)-resistant ties for use outdoors.

- ✔ **Electrical tape:** Use high-quality tape (I like Scotch 33+) for important jobs, such as sealing outdoor connecters, and stock up on bargain tape for temporary or throwaway jobs. Watch for sales on tape of all sorts at home improvement and hardware stores.

- ✔ **Filters:** Having on hand a couple of broadcast or TV filters, ac power cord filters, and *ferrite cores* (see Chapter 18) allows you to address interference problems quickly.

- ✔ **Fuses:** Have spares for all the fuse sizes and styles your equipment uses.

- ✔ **Metal fasteners:** If you're not already well stocked, purchase a parts-cabinet assortment with #4 through #10 screws, nuts, and lock washers. Some equipment may require the smaller metric-sized fasteners. For antennas and masts, ¼-inch and ⁵⁄₁₆-inch hardware is common.

- ✔ **Solder:** For electronics work, use a *rosin flux core solder. Never* use *acid core solder* or *acid flux,* which are intended for plumbing work. The acid corrodes electronics quickly and horribly. The type of flux in solder is printed prominently on the label. For general repair and maintenance, solder with a diameter of around 0.032 inch is about right. If you need to get guidance on how to solder, *Electronics For Dummies* has a good guide and www.mediacollege.com/misc/solder is a good online tutorial.

If you're not sure what you're doing, ask a more experienced friend to help you with the soldering.

The Mechanics of Stocking Spare Parts over Time

Although having a complete inventory now would be nice, it's not necessary; you can build up a stock of the kinds of components you need over time. Rather than give you a huge shopping list, here are some guidelines to follow:

- ✔ **Buy more than you need:** When you buy or order components for a project, order extras. Electronic components, such as resistors, capacitors, transistors, diodes, and so on, are often cheaper if you buy in quantities of ten or more. After a few projects, you'll have a nice collection.

✔ **Shop the sale aisle:** Watch for sales and monthly specials at your favorite vendors. You can sometimes get small quantities of parts at large quantity pricing.

Avoid so-called *grab bags*. These are often full of junk or parts with a questionable pedigree.

✔ **Buy in bulk:** Small hardware components (screws, nuts, and washers) are always cheaper in quantity. If you buy a box rather than the little plastic bags, you'll never have to make a run to the store in the middle of a project again!

✔ **Enter the ham zone:** The electronic flea markets known as *hamfests* (see Chapter 9) are excellent sources of parts and component bargains. Switches and other complex parts are particularly good deals. Parts drawers and cabinets often come with parts in them and you can use both!

If you decide to get really serious about the electronics of radio, you'll need to have a stock of electronic components. Table 16-2 lists common components that I find myself using in most of my electronics projects. These are useful for building new projects or for doing repairs. You don't have to buy them all at once, but when you get an opportunity, these are good ones to have around.

You can save yourself a bit of shopping by picking up component assortments that often come in handy parts cabinets. Jameco Electronics (www. jameco.com) is a good source for inexpensive resistor and capacitor assortments.

Table 16-2	Commonly Used Electronic Components
Part	*Common Types*
Resistors	Various values of 5% metal- or carbon-film, ¼- and ½-watt; 100, 500, 1k, 5k, 10k, 100k variable resistors
Capacitors	0.001, 0.01, 0.1 (F ceramic; 1, 10, 100 (F tantalum or electrolytic; 1000 and 10000 (F electrolytic; miscellaneous values between 220 and 10000 pF film or ceramic
Inductors	1, 10, 100 mH chokes
Semiconductors	1N4148, 1N4001, 1N4007, and full-wave bridge rectifiers; 2N2222, 2N3904 and 2N3906 switching transistors; 2N7000 and IRF510 FETs; red and green LEDs
Integrated circuits	7805, 7812, 7912, 78L05, 78L12 regulators; LM741, LM358, LM324 op-amps; LM555 timer; LM386 audio amplifier; MAX232 RS232 interface

Your friend, the junk box

A well-stocked junk box is one of the biggest time and money savers and every true-blue radio guru has at least one. (I count four in my shop.) Start with one and work your way up to true Radio Bliss! Old paint trays and dishpans make rooting through the heap in search of that special part a bit easier. Junk boxes come in several flavors:

✔ **The hardware junk box:** Build up a hardware junk box by tossing in any loose screws, nuts, spacers, springs, and so on that you come across. Someday, that part will be just right for a repair job, finishing a construction project, helping out on a science fair project, or repairing something for your tolerant spouse.

✔ **The electronics junk box:** Use this junk box to hold useful pieces of broken appliances and entertainment devices. Strip your dead appliances of the good stuff and throw away the carcasses. Transformers, headphone and speaker jacks, switches, fuses,

and lots and lots of interesting hardware can be gleaned that would otherwise end up in the dump. Plus, taking apart those old gadgets is interesting — you get to see how they're made.

✔ **The wire and cable junk box:** I also recommend starting a wire and cable junk box. That 2-foot piece of house wiring contains the materials for a fine scanner antenna, for example. The power cord salvaged from the table lamp the cat broke (at least, that's what you *said* happened) may be just the ticket for repairing an antique appliance.

✔ **The "what the heck is this?" junk box:** Use this junk box to house all those odd plastic parts, hybridized stuff, broken bits, and anything you can't quite figure out but aren't ready to pitch. Save the extra hardware and plastic doodads that come with an appliance or gadget. These parts may come in mighty handy for repairs.

Finding Education and Training

The most valuable tool in your kit isn't anything you use with your hands; it's what's between your ears! You already know that because you went out and bought this book to satisfy your curiosity about radios. The process of learning is lifelong and the more you discover, the more you want to discover. An educated radio user gets more out of the radio and has more successful communications.

Lucky for you, you can find lots and lots of resources — tutorials, guides, courses, and FAQs (Frequently Asked Questions) — out there that you can take advantage of at a minimum cost.

Getting freebies from manufacturers and retailers

The first place you should turn for free information is the Support and Downloads sections of the Web sites of radio manufacturers. The manufacturers know that a frustrated customer is an unhappy customer. It's in their interest to provide information about operating radios, properly installing components and add-ons, and using antennas. Manufacturers also want users to have reasonable expectations about how the radio is likely to perform. The more you know, the less likely you are to be frustrated and the more likely you are to be a repeat customer.

The sales staff and technicians at radio shops are also a wealth of knowledge there for your taking. They are especially knowledgeable about the local and regional conditions that might affect how your radio works. Talk to these people about the hot channels and find out about local events. They may even be able to provide frequency lists to help you get started. If you are confused about how the controls on your radio work or about the value of a certain function, these are the folks to ask.

Visiting individual or club Web sites

One of the oldest traditions in radio is to assist newcomers. That tradition is still going strong! There are many electronics and radio tutorials and reference Web sites available at a click of the mouse. A single author or organization develops some of these, while some are more like portals that collect items and information for a variety of sources.

You should *always* be a little wary of the information you find on the Internet, of course, because you can't be sure of an author's background or expertise. Organizations tend to present more reliable information than individuals do. It's a good idea to acquire information from a couple of different sites on the same topic. That way, you'll get a couple of different perspectives on the same material. A serious conflict in information is a clue that perhaps one of the sites is a little off the beam and another reference is needed. Table 16-3 provides a few sites that I think are useful.

Getting information from books

Obviously, you understand the value of a good book. Good thing because self-study is a time-honored way of educating yourself about electronics and radio.

When shopping around, remember that although a reference book need not be brand new, it should be printed in the past few years to include a good helping of new technologies.

Table 16-3	Electronics and Tutorial Sites
Web Address	**Subjects and Type of Coverage**
www.discovercircuits.com	A collection of circuits and information run by David Johnson, an electronics engineer
www.electronics-tutorials.com	Assorted tutorials on radio and electronics by Ian Purdie, an Australian radio ham with the call sign VK2TIP
dmoz.org/Science/Technology/ Electronics/Tutorials	Links to a number of tutorial sites covering all kinds of electronics

Getting basic training

These books are aimed at the beginner who wants to start at (or be refreshed on) the ground floor. In order of most basic to most challenging, here are my picks for basic training in electronics:

- ✔ ***Guide to Understanding Electricity and Electronics*, 2nd Edition, by G. Randy Slone, (McGraw-Hill):** If you're familiar with the basic electronic concepts and want to move to more advanced electronic circuits, this is an excellent book. It's not too "gee-whiz" in nature, nor is it a textbook. The book takes the reader from elementary circuits through simple transistor and amplifier circuits. Radio and computer electronics come in towards the end of the book.

- ✔ ***Teach Yourself Electricity and Electronics,* by Stan Gibilisco, (TAB/ McGraw-Hill):** This book is organized as a series of tutorials organized as if the student was taking a class, with each lesson building on the one that precedes it. The book assumes that readers have some background, covering many topics starting with ac and dc electricity before moving to more advanced topics.

- ✔ ***Understanding Basic Electronics,* by Larry Wolfgang (American Radio Relay League):** Also aimed at the beginner, the book starts with some practice in the required math and then progresses through electricity and electronics. It offers a series of tutorials and tabletop experiments.

Getting reference guides

If you're looking for a good reference book that you can pull off the shelf whenever you have a question, the following list is a good place to start:

- ✔ *The ARRL Handbook,* **2005 Edition, edited by Dana Reed (American Radio Relay League):** Without a doubt, the reference that every radio user should have on the shelf is *The ARRL Handbook* (www.arrl.org). The 2005 edition is the 82nd edition. It is referred to as simply *the handbook* by many radio enthusiasts. It is the Bible for serious radio users. It has more radio know-how packed into its pages for the price than any other book around. It also comes with a CD of software for various radio design chores. Although the book is targeted at the ham radio audience, most of the information is just as applicable to any other radio service. Radio is radio!

- ✔ *The Black & Decker Complete Guide to Home Wiring* **(Creative Publishing International):** Although this isn't an electronics and radio book, you will likely wind up needing to run a new antenna cable, install a new circuit breaker, ground an antenna mast, and so on. When you do, it is a good idea to do it properly and safely. This book shows you how.

- ✔ *Electronics For Dummies,* **by Gordon McComb and Earl Boyson (Wiley):** This is a beginning-level book about very basic electronics and information about the common devices and components. It also includes examples of construction techniques.

- ✔ *Practical Antenna Handbook,* **by Joe Carr (TAB/McGraw-Hill):** Joe covers antennas of all sorts, from low-frequency wire antennas to serious microwave antennas. He also addresses mobile and marine antennas.

Taking online courses

You can find a ton of local schools and colleges that offer online training and education. If you're willing to pay for these classes, go for it. If (as I suspect) radio is still a hobby and you're not really interested in getting a degree or certificate in electronics, you may want to think twice before you commit your money and your time.

If you would like to take courses in electronics and radio for free or at a low cost, here are a couple of good resources for you:

- ✔ **The American Radio Relay League's Certification and Continuing Education program:** ARRL (www.arrl.org/cce) offers a number of low-cost courses with discounts for members. Although they're geared to the radio ham, the information is applicable to radio enthusiasts in general. Your friendly author has written several courses, including Analog Electronics (course number EC-012) and Digital Electronics (EC-013). These are mentored courses, with exercises and construction projects for each of the 16 modules.

✔ **U.S. Navy Electricity and Electronics Training Series (NEETS) self-study course:** This self-study course on electronics and communications is available at www.phy.davidson.edu/instrumentation/NEETS.htm. You can download the information in PDF format. Each of the 24 modules is at least 100 pages long, so it's a meaty course. This is complete and thorough information. Don't worry. You won't have to whistle "Anchors Aweigh" while you work. You can study the modules in any order and at any pace.

Chapter 17

A Spark of Electronic Know-How

*Y*ou can operate a radio without knowing a single thing about electronics if you're good at following instructions and diagrams. However, when you try to do something the manual doesn't cover, you will be well served to know a little bit about basic electricity. Installing your radio properly requires some understanding of basic electricity, too. Although this chapter won't make you an electronics wizard, it does explain some of the terms you'll encounter.

Understanding the Relationship between Amps, Volts, Watts, and Ohms

My goodness, more scientists! Meet Professors Ampere, Volta, Watt, and Ohm. (I once rented a house from the great-granddaughter of Professor Ohm, believe it or not!) From the work of these electrical pioneers, we have the unit of current (amperes, amps, A or I), voltage (volts or V), power (watts or W), and resistance (ohms or Ω). What do all these words mean? The following sections give you the information you need to understand what they measure and how they relate with each other.

All the jargon for defining and measuring electricity can be broken down into a few crucial topics:

✔ Current, measured in *amperes* (the abbreviation is *amps*), defines the rate of flow of electricity.

✔ Voltage, measured in *volts,* is the electrical force that makes current flow between two points.

✔ Power, measured in *watts,* is the rate of energy generation or consumption

✔ Resistance, measured in *ohms,* defines how much current will flow in response to an applied voltage.

Measuring current

You need to understand how *currents* work because your radio draws current when it's operating. Knowing about current can help you troubleshoot and ensure that your radio's power source and wiring can adequately (and safely) handle a given current. Similarly, if you are using a battery-powered radio, you need to choose a battery capable of supplying enough current to run the radio.

When you turn on the faucet and water squirts out (assuming you've paid your water bill), you judge the flow by how quickly a volume of water is produced, filling your bucket or pan. Electrical current, like water flow, is a measure of how many units of electrical *charge* pass through a connection or wire per second.

The unit of electrical charge is a very small particle called an *electron.* How small? Is subatomic small enough for you? When you flip the switch of that 100-watt light bulb in your kitchen, one ampere, or *amp,* of electrical current flows through the bulb. That one amp of current is composed of 6.24 billion billion (yes, billion *billion*) electrons flowing through the bulb every second!

Electrical current is a measure of how many electrons are being pushed through a wire, just as water flow is a measure of how many gallons of water are being pushed through a pipe. Electrical current is always measured at a specific point and it flows *through* something. For example, you might hear yourself saying, "There are 2 amps of current flowing through the red wire."

A specific amount of current is specified in amperes or amps, abbreviated A, but a general reference to current without a known value is abbreviated I. I and A mean pretty much the same thing, but the convention is to use A when you know the exact value of amps and I when you don't. (Hey, if it was made too easy, everybody would be wearing a lab coat and doing experiments!)

Any electrical path through which current can flow is called a *circuit.* If the path is interrupted, it becomes an *open circuit.* If a circuit is created that bypasses the intended circuit, the new circuit is called a *short circuit*, especially if the result is a malfunctioning circuit.

Understanding voltage basics

Electrical currents are made up of electrons moving from a power source through some sort of conduit, such as a wire. So what's making these electrons move? Just as there must be pressure in a pipe to make water flow when you turn on the faucet, there must be electrical pressure to make the electrons flow in a wire, creating current.

Electrical pressure, called *potential,* is measured in *volts,* abbreviated with the familiar V. Voltage is measured *between* two points (from Point A to Point B). You can make a direct correlation between electrical pressure and water pressure, which is also always measured between two points.

One of the two points is usually a *reference pressure.* For electricity, the reference pressure is usually called *ground* or *common* and is the reference to which all voltages are measured in the equipment being discussed. For example, if I say, "I'm measuring 10 volts," I have to tell you between what two points I'm measuring these 10 volts or you have to assume that one of the points is the common reference.

It's interesting to note that just as no water will flow in a pipe connected between two sources of water at the same pressure, no electrical current will flow in a wire connected between two points at the same voltage. Figure 17-1 illustrates the analogy.

Although V is the most common abbreviation for voltage and will be used exclusively in this book, you may also encounter E, which means pretty much the same thing. The E abbreviation refers to an alternate term for voltage called *electro-motive force* (the thing that makes electrons move). You may also see the abbreviation lower-cased (e or v) in other books and articles.

Current flows *through* a specific path while voltage exists *between* two points along the path. Voltage doesn't flow through anything, current does. Voltage is simply a measure of the pressure that makes current flow.

Voltage is important to radio users because the power source you use to run the radio must have the correct voltage. The power source must also be able to supply all the current the radio needs without letting the voltage drop in value. Batteries have a rated voltage and their size indicates how much current they can supply. Batteries are discussed in Chapter 19.

Calculating power

The power delivered by an electrical current is determined by calculating the pressure and flow in the form of voltage and current. As a correlation, when you talk about the amount of power available from a water system, you have to specify both pressure and flow. Well, the same is true for electrical systems. Electrical power, P, is measured in watts, and is equal to voltage multiplied by the current, or V x I. For example, if a 6 V battery is connected to a motor that draws 3 A, the motor is using 6 x 3 = 18 watts of power.

Figure 17-1:
Electrical
voltage and
current act
similarly to
pressure
and flow in
water
systems.

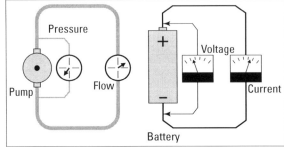

A stands for the units of amperes, and I stands for an unknown current when you're making some sort of calculation. You can calculate voltage (V), current (I), or power (P) if you know the other two quantities by following these simple equations:

Power in watts (P) = voltage in volts (V) × current in amps (I)

```
P = V × I
```

Voltage (V) = power (P) ÷ current (I)

```
V = P ÷ I
```

Current (I) = power (P) ÷ voltage (V)

```
I = P ÷ V
```

For example, a 40-watt light bulb with 120 V across it has a current of `40 ÷ 120 = 1/3` `A` flowing through it. A 1000-watt heater drawing 8 A must have a voltage of `1,000 ÷ 8 = 125` `V` running across it.

You'll encounter power often when you use your radio. How many watts does a radio consume? How many watts does it produce? Are your cables and antennas adequately rated to accommodate your radio? By knowing how to calculate power, you can choose the right parts to put your radio on the air.

Introducing resistance

When you water your garden, you don't just turn on the water full blast; to limit flow you open the valve part way, use an adjustable nozzle, or place your thumb over the end of the hose. All three of those methods are equivalent to placing *resistance* in the path of an electrical current. Just as an adjustable valve allows you to control water flow, electrical resistance allows you to control current. For a given amount of water pressure, the resistance of the valve, nozzle, or thumb determines water flow. Higher resistance means less flow. Similarly, for a given amount of resistance, higher pressure means higher flow.

Professor Georg Ohm discovered the relationship between voltage *(V),* current *(I),* and resistance *(R).* He figured out that the voltage across a conductor is proportional to the current that travels through it. This relationship is named after him — *Ohm's law.* The unit of measure for resistance is the *ohm,* abbreviated with the Greek omega symbol, Ω.

Electrical resistance is defined as the voltage divided by the current that flows as a result, or R = V ÷ I. If a high voltage is applied, but little current flows (imagine a clogged pipe); then resistance is high.

Resistance is present in all electrical connections and converts some of the current's electrical power to heat, just as friction consumes some of the power of water flowing in a pipe. *Resistors* are electrical components used to limit or control current in electronic circuits. They have a fixed amount of resistance and can *dissipate* a certain amount of heat. Some typical resistors are shown in Figure 17-2.

Making calculations with Ohm's law

By using Ohm's law, which states that voltage is proportional to current, you can determine resistance, voltage, or current. If you know any two of the remaining three quantities, the calculations are fairly simple:

Resistance in ohms (R) = voltage (V) ÷ by current (I)

```
V ÷ I = R
```

Voltage (V) = resistance (R) × current (I)

```
R × I = V
```

Current (I) = voltage (V) ÷ by resistance (R)

```
V ÷ R = I
```

For example, 10 amps flowing through 5 ohms of resistance results in `10 × 5 = 50 V` across the resistance. If the 1,000-watt heating element draws 8 A with 125 V across it, its resistance must be `125 ÷ 8 = 15.4 Ω`.

Adjustable potentiometers (pots)

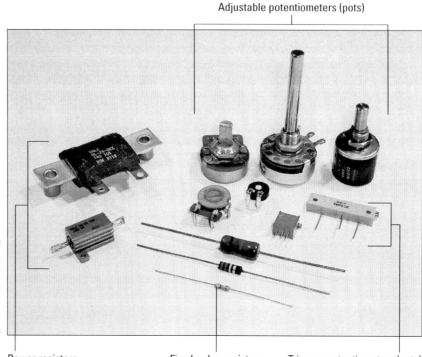

Figure 17-2:
Examples of
different
types of
resistors.

Power resistors   Fixed-value resistors   Trimmer potentiometers (pots)

Because resistors turn electrical power into heat, you may occasionally need to know how much heat is produced. In order to calculate how much heat is produced, you need to know the resistance (in ohms) and either the voltage across or current through the resistor.

Power in watts (W) = voltage (V)_ ÷ by resistance (R)

```
V² ÷ R = W
```

Power in watts (W) = Current (I)_ × resistance (R)

```
I² × R = W
```

For example, applying 12 V to a 100-ohm resistor results in `122 × 100 = 144 × 100 = 1.44 W` of heat.

If 100 milliamps or mA (that's 0.1 A) are flowing in a 2-ohm resistor, the heat produced is `0.12 × 2 = 0.01 × 2 = 0.02 W = 20 mW` (milliwatts).

Although you may not have to deal with resistance very frequently, it's important that you understand what it is. Choosing the right size of wire or connecter depends on you understanding how resistance affects electrical current, voltage, and heating. Even if you don't have to perform detailed calculations, a super-hot wire or a low voltage can cause all kinds of problems. Knowing how to measure resistance helps you troubleshoot your radio system, as well.

When ac current flows through a circuit containing inductors (coils) or capacitors, it encounters not only resistance but also *reactance*. The combination of resistance and reactance is called *impedance*. You'll encounter this term when dealing with radio feedlines and antennas. All you need to remember is that impedance is to ac current what resistance is to dc current.

Wires, Cables, and Connecters

A great story is one of an elderly lady being shown her first radio back in the early 20th century. Her response was, "I don't know why they call it wireless! I've never seen so many wires in all my life!" It's true, for a wireless technology, even 21st century radios sure require a lot of them.

Wires

As shown in Figure 17-3, a *wire* has only one electrical *conductor*, made of metal and either left bare or covered with a coating of *insulation*. The conductor can either be a single strand, which is referred to as *solid wire*, or multiple strands, called *stranded wire*. Bare wire (without any insulation) is mostly used for grounding (where the voltage on the wire is zero) or for antennas, where the wire is up in the air somewhere and can't be easily touched. Insulation is used whenever more than one wire carrying different voltages is bundled together (such as for house wiring).

A wire's size is specified by its *gauge*. The lower the gauge is, the thicker the wire. Thicker wire can handle more current because it has lower resistance, just as a larger pipe can handle a larger flow of water.

For use in powering your radio, you should be primarily concerned with the wire's current-carrying ability. Antenna wire is chosen mostly on the basis of its ability to withstand the elements. If you decide to do some wiring inside a piece of equipment, use *stranded hookup wire*. It's small and flexible.

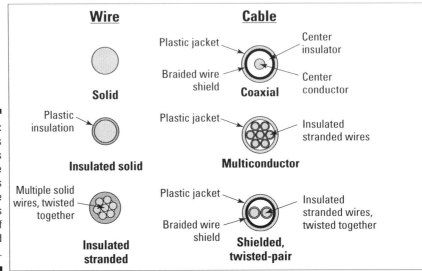

Figure 17-3: Cross sections show the differences between the various types of wire and cable.

Cables and cords

When more than one wire is packaged together inside an outer sleeve, or *jacket,* the package is referred to as *cable,* to distinguish it from the individual wires it contains. There are many different types of cable (refer to Figure 17-3). Cables that just contain two or more wires inside an outer jacket are called *multiconductor cables.* Sometimes, the wires are twisted together in pairs and, strangely enough, these cables are referred to as *twisted-pair* cables.

Cables may also have a flexible sleeve of braided wire, called a *shield,* around some or all the wires and inside the jacket. These are appropriately called *shielded* cables; the shield keeps stray radio signals from contaminating the signals being carried by the wires inside the cable.

The cable you will deal with most is called *coaxial cable,* or simply *coax.* Coax is a special type of cable made especially for carrying radio signals. Coax has a central conductor that may be solid or stranded, surrounded by a thick layer of solid or foam plastic called the *center insulator.* The next layer is a shield. Coax shields are typically heavier and have fewer holes because coax is designed to keep radio signals moving from one point to another and not let them *leak* out or in. A plastic jacket then covers the shield. The cable's name comes from all the conductors sharing a common central axis, or *co-axial.*

A *cord*, such as a *patch cord* that carries audio signals or *power cord* that connects equipment to the household ac wall socket, is a cable that has had connecters attached to it. If a short cord is made of coaxial cable for connecting pieces of equipment, it is referred to as a *jumper*. (These are not to be confused with jumper cables that you use for starting a car with a dead battery.)

Connecters

This section describes the various types of common connecters that you'll encounter.

Here is some common connecter terminology you should know:

- A *plug* is a connecter on the end of a cable.
- A *jack* or *receptacle* is the mating connecter mounted on a piece of equipment.

  Plugs, jacks, and receptacles generally have at least two *contacts* that connect to separate wires carrying signals within a cable.
- A *pin* is a single contact that is inserted into a hollow contact called a *socket*.

  Plugs and jacks can have all pins, all sockets, or a mix of both.
- A *shell* or *backshell* is the part of the connecter that protects the inner contacts. The shell may also have a *strain relief* that clamps the wires or cable to prevent flexing.
- A *male connecter* is a connecter in which the signal contacts are exposed pins (disregarding the shell).
- A *female connecter* has sockets or receptacles that mate with the pins.

Direct current connecters

Terminal blocks, barrier strips, Molex connecters, coaxial plugs, and miniature phone plugs are all shown in Figure 17-4. Here's some information about these connecters:

- To use connecters in a *terminal block*, individual bare wires are inserted into a common hole and a screw then clamps them together.
- A *barrier strip* has opposing pairs of screws, connected together, but isolated from the neighboring pairs. Either bare wires or crimp-on terminals can be used with the barrier strip.

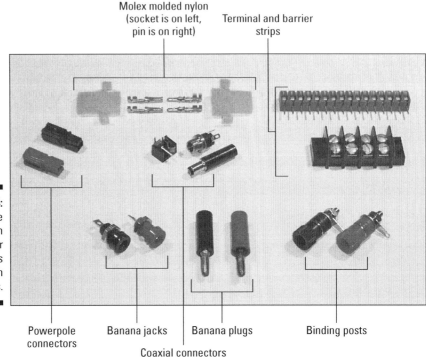

Molex molded nylon
(socket is on left,
pin is on right)

Terminal and barrier
strips

Figure 17-4:
These are
common
power
connecters
used with
radios.

Powerpole
connectors

Banana jacks

Banana plugs

Binding posts

Coaxial connectors

✔ The *Molex connecter* is a brand name of the Molex Company. The body of the nylon connecter accepts special pins and sockets that are crimped onto wires and inserted into the connecter.

✔ *Powerpole connectors,* dc power connectors made by the Anderson Company, are rapidly becoming a standard power connecter. They are *genderless* connecters because they have no male and female versions.

✔ The *coaxial power connecter* has nothing to do with coaxial cable, but is so-named because both contacts, the sleeve and the tip, are round and share the same central axis.

✔ The *miniature phone plug* is just like the ones on your portable stereo headphones, but with only two contacts as is sometimes used for low-power dc connections.

Understanding why coaxial cable has impedance

When you go to the radio store to buy some coax cable, you have to choose between types of cable that have impedances of 50 or 75 ohms. You may see *twin lead* cable for TV antennas with an impedance of 300 ohms. Yet, if you whip out your handy voltmeter and measure the resistance of the cable from end to end, you'll discover a resistance of a fraction of an ohm. What's going on? The voltmeter only uses dc current to measure resistance of the conductors and doesn't see the effect of the cable on ac current.

The physical construction of coax cable causes it to have a certain impedance that affects its ability to carry RF signals. In analogy, consider two straws; one with a wide cross section and the other with a narrow cross section, like a coffee stirrer. If you try to blow a short puff of air

through each, the wide one will resist your breath much less — it has a lower physical impedance to changes in airflow. The narrow straw has a higher impedance. The electrical impedance of coaxial cable doesn't depend strictly on diameter — there are skinny cables and fat cables with the same impedance — but on the materials used to make it, as well as the arrangement of those materials. Radio systems work best when the connections between the components have matching impedances. In other words, the transmitter, cable, and antenna should all have similar impedances. The standard impedance used for radio is 50 ohms, while video systems prefer 75 ohms. Use cable with an impedance that matches the equipment's specification in the operating manual.

Audio and control signal connecters

Audio and control signal cables and cords usually have one of these connecter types:

- ✔ The *phone plug* got its name from being installed on the cables used by old-time telephone operators to make connections. The contact at the end of the plug is called the *tip*. The long metal tube between the phone plug's tip and the shell is called the *sleeve*. If there is a short contact between the tip and sleeve, it's called the *ring*. The *standard* phone plug's sleeve contact is ¼-inch in diameter. There are two smaller phone plugs; *miniature* (⅛-inch or 3.5 mm) and *subminiature* (³⁄₃₂-inch or 2.5 mm). If the phone plug or jack only has two contacts, it's referred to as *mono* (because it can only carry one audio signal) and if it has three contacts, it's *stereo*.

- ✔ *Phono plugs* or *RCA plugs* only have a center pin and outer shell contact. They were originally popularized by RCA to connect phonographs to amplifiers — thus the name. Phono plugs can also be found carrying RF signals at low power.

✔ *DIN* comes from *Deutsches Institut für Normung,* an international standards organization that has certified several styles of multipin connecters. These connecters are popular on microphone cables (particularly the 8-pin, round style) and for accessory or data connections (usually the 5- or 7-pin computer accessory style).

Modular connecters (such as the RJ-45 modular connecter used for computer networks) have become popular as microphone connecters, as well. Although you can install all the other connecters by soldering, the modular connecter requires special cable preparation and crimping tools.

RF connecters

When it comes to connections carrying the RF signals to and from your radio, special connecters are required. These ensure that all the precious RF is transferred properly because these high-frequency signals are sensitive to impedance and require high-quality connections to avoid signal loss.

To install these connecters requires some practice, so I recommend that you start with premanufactured cables or have a knowledgeable friend install them for you while you watch carefully.

Here are some common RF connecters:

✔ The *UHF* connecters are the usual connecter for CB, shortwave, and ham radios.

✔ The *PL-259* is the cable-mounted male plug and the *SO-239* is the female receptacle mounted on the equipment.

✔ The double-female PL-258 or *barrel* connecter allows two cables to be connected together. A threaded shell holds the connecters together.

✔ *BNC* plugs and jacks are the standard on most scanner and handheld radios. Make a BNC connection by pushing the plug down onto the jack and twisting the shell of the plug until the pins on the shell snap into place.

✔ *F-type* plugs and jacks are found on TV and video equipment and cables. The threaded outer shell holds the connecters together in a similar fashion to UHF connecters.

The center contact of F–type plugs is just the solid center conductor of the coaxial cable. These are very inexpensive plugs, but work very well if properly installed.

You may also encounter other less-common connecter types, such as the *N* and the *SMA,* which are used where special techniques are required to maintain good connections. Sometimes manufacturers create their own custom connecters. If that's the case with your radio, you have no choice but to use the cables and accessories provided by the manufacturer. In general, I recommend that you avoid purchasing a radio that uses *proprietary* connecters, cords, and accessories.

Schematic diagrams

When you have more radio experience under your belt, you'll want to become familiar with how to read electronic circuit diagrams, also called *schematics*. Schematics provide a description of how equipment and components are connected together. To master this skill, you simply need to keep practicing. It's not necessary to read schematics in order to get your radio working, but it will open up new avenues of understanding and make interacting with other radio enthusiasts a lot easier. Don't shy away from them.

Adapters

Adapters are used to make a transition between types of connectors or to create unusual sets of connections. For example, the PL-258 barrel connecter mentioned earlier is a type of adapter. If you want to connect a cable with a PL-259 plug on it to a piece of equipment with a BNC receptacle, you would need a male-UHF-to-female-BNC adapter. Because there are many types of connecters, there are even more types of adapters. There are also adapters that allow multiple connecters to make contact with a single receptacle or that reduce or increase the number of contacts, such as a mono-to-stereo converter.

Many radios today have the ability to communicate with a computer. I discuss computer connecters in Chapter 20.

Dealing with Safety Issues

A few words on electrical and RF safety are always in order. As you are aware, household ac voltages present serious shock hazards. Never work on a piece of equipment that is connected to ac power unless you are an experienced technician. Always use proper line cords that are in good condition. Throw away those frayed cords with bent plugs. If the equipment has a ground terminal on its metal case or *chassis*, be sure it is connected to a safety ground, either through a three-wire power cord or with a wire connected to a ground rod.

Even though dc voltages used by radios are generally much lower than the household ac voltage, the higher currents can present hazards of their own. Large batteries can supply enough current to melt wires and start fires.

Be sure that connections on dc-powered radios are fused in both the positive and negative leads. This is a particular concern when installing radios in cars because vibration and the tough automotive environment can shake connections loose or cause a short circuit. Disconnect one terminal of your battery whenever you are working on a dc power system at home or in your car. (I learned this the hard way!)

Chapter 18

Installing Radios Right

. . . .

. . . .

*E*very installation has a different set of circumstances, so I can't give you step-by-step instructions that match *every* scenario. What I can do in this chapter is give you good ideas about the things that you need to think about before you go around drilling holes or making your bid to take over that unused corner of the spare bedroom for your radio needs. I start out by nagging you about safety issues, and then get to the fun part, showing you how to set up your radio at home and in your vehicle. I also give you information about setting up additional connectors.

Installing Your Radio at Home

There are as many different home radio *shacks* as there are homes. Ranging from a corner of a hallway table to the entire basement, no two radio home bases are the same. Look closely, however, and you will see that all shacks have common elements. The following sections cover all the common elements so that you can take everything into consideration. If you're a business owner planning on using a radio, you can consider the same issues from a workplace perspective.

Setting up a safe radio environment

I'm sure you're looking at that shiny radio thinking, "How on earth could there be any safety issues?" With the low-power radios and no-power scanners and receivers, what could go wrong . . . go wrong . . . go wrong?

Regular power safety issues

Unless the radios and gadgets that go with them are running solely on battery power, they need some kind of external power supply. Usually, this means a *wall wart* power supply (also known as an ac adapter, wall transformer, or power pack) that plugs directly into the wall.

Wall warts pose very few safety issues — unless you try to plug too many of them into one outlet. Don't give in to the temptation to jam all of your plugs into that same hapless extension cord. Especially, don't cut off the safety ground pin (check out the section called, "Grounding," later in this chapter)!

If you're using *power strips* (see Chapter 21), no matter how much or how little you pay (they can be inexpensive or very pricy), you should always have a couple on hand. Attach the power strips to your radio table or desk for convenient, reliable use.

Every once in a while, check each power supply to see if it's getting hot — an indication of impending failure.

Advanced power safety issues

You may choose to have a separate dc power supply if you have a CB radio or several items that operate from dc power. Most radios require 12 V.

Here's why you need to be careful when you set up your dc power system. At first, you have one or two power cords connected. Then three. Then more, and the next thing you know, you have an ungainly and unsafe mess under your desk. When this happens, it's time for a *power distribution strip*, such as the RIGrunner (www.westmountainradio.com) or the *outlet strips* made by MFJ Enterprises (www.mfjenterprises.com). Power distribution strips make it easy to provide power to many different pieces of equipment without the usual rat's nest of trouble waiting to happen.

I cover batteries thoroughly in Chapter 19. For now, I just remind you that they store a lot of energy inside, so you need to be careful not to let them get shorted out — even the small ones — unless you like hot wires. Don't overcharge them — they get hot or corrode!

Grounding

Because you're not operating a high-power radio, you need to be concerned primarily with what is referred to as a *safety ground*. A safety ground is an alternate current path whose primary purpose is to conduct unwanted current away from the equipment user, hopefully blowing a fuse or tripping a breaker at the same time.

Good grounding protects you against shock and your equipment from stray radio signals getting into connections that they shouldn't get into. In general, if your equipment comes with a three-wire ac cord, be sure the safety ground pin (the round one in the U.S.) stays connected to the ground pin of your home's ac wiring. This would be a good time to check the outlet wiring with an *outlet tester*, such as the RadioShack 22-141.

If your radio has a metal enclosure or *chassis*, and operates from a dc power supply or two-wire ac cord, check the operating manual to see if the manufacturer recommends attaching a ground wire. The ground wire can be connected to your home wiring safety ground by clamping it under the center screw of the ac socket cover plate or to a separate ground rod.

If you're setting up radios for use by employees or members of the public, be sure to have your installation performed or inspected by a licensed electrician.

Setting up your very own radio central

Before you set up your radio, consider what you intend the radio to do for you. Provide emergency communications? Act as a family intercom? Be your personal hobby? You should consider the following issues:

- ✔ Location
- ✔ Comfort
- ✔ Safety

Finding a location that is easily accessible but not distracting to others

For use by multiple people, such as in emergencies and day-to-day communication, any and all radios you set up need to be in convenient, accessible locations, as shown in Figure 18-1. On the other hand, they shouldn't be in the way or in disruptive spots. Many home users find that they already have a *communications central* near an answering machine or cordless phone base. Another good spot can be in the kitchen or receptionist's area, particularly if you plan to keep the radio on all the time, listening for calls. Whatever location you choose, gather the radio equipment, including manuals, chargers, and spare batteries.

If you're using your radio for the entertainment or recreation of one person, consider putting the radio central out of the way of other family members. Continuous radio chatter is an irritating distraction to other folks, even if it *is* fascinating to you.

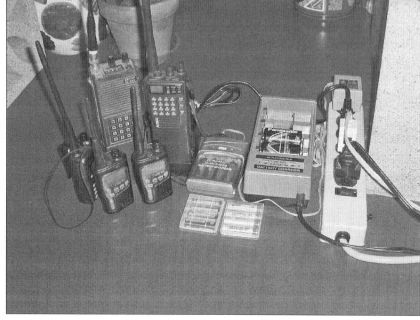

Figure 18-1:
Create your own radio central where you keep all handheld radios, batteries, and the battery charger.

If you can't locate the radio away from the rest of the family, invest in a set of headphones. See the section, later in this chapter, called "Headphones, communications combos, and speakers" for more information.

Finding a place that is comfortable for the user

Spending hours listening to a radio such as a shortwave receiver or scanner is pretty common. You'll quickly find that you need to consider the listener's convenience and comfort. Avoid cold or hot rooms and shelves too high or low to be comfortably reached, if the radio is going to be frequently tuned or adjusted. Lighting should be indirect and glare free. You should set up the lighting so that it is behind the operator, allowing users to easily read the various dials and displays. Figure 18-2 shows one possible setup. Note that the rolling cabinet is inexpensive, but it has room for an antenna rotator control box and battery backup.

Finding a location that keeps power cords out of the way

Power cords are unsightly, but more important, they get in the way and quickly become an aggravation — or even a hazard. Power supplies are best placed behind or below the operating position. If you have a lot of cables, you can keep them neat with *cable clamps,* such as the reusable CableClamp (www.cableclamp.com/), Velcro strips, or plastic-covered wire ties from the grocery store. In my experience, keeping cables under control means fewer accidents and makes the station a whole lot easier on the eyes.

Figure 18-2:
A listening station with a ham radio, CB, and scanner sitting on top of a rolling cabinet.

To keep your radios from flying about in an earthquake, use nylon webbing to secure the radio's mobile mount to a shelf or desktop. The radio can then be secured to the mobile mount, allowing it to be easily removed and maintained.

Accessories

After the radio makes its appearance, the parade of accessories won't be far behind. It starts with just a couple of audio gadgets, such as headphones and a desk microphone, perhaps. Then the spare batteries and charger appear. If you're a CB operator, you may find yourself buying a radio frequency (RF) power meter, as well. You can say it won't happen to you, that you can control yourself, but you're wrong.

Allocate space *now* for the useful items you'll bring in with your radio, and don't forget to make room for the books, magazines, and printouts you'll find yourself accumulating.

Headphones, communications combos, and speakers

Headphones designed for a personal stereo will do if they're comfortable and of good quality. You can get headphones in a couple of styles. The open-air styles with foam pads allow you to hear what's going on around you. If you would rather block out the noise and hubbub around you, earphones are available with heavy ear cups (called *closed cups*) to keep unwanted noise out.

You can also get special communications headsets that include a built-in microphone. Communications *headsets* have a *boom microphone* on an extension that's placed near your mouth.

You can also get a spare set of headphones and a *splitter* so that two people can listen at the same time. Or you can attach a recorder to the second line.

Your radio probably has a speaker output, too. If you connect a good computer or communications speaker, you may be surprised at how much better the audio sounds than on the built-in speaker. Good quality headphones and speakers can substantially increase the intelligibility of signals, reducing operator fatigue, as well.

Microphones

On the operator end of the transmitter comes the microphone or *mic*, a popular item to experiment with. Remember that a cheap mic is not going to make you sound good on the other end. The microphone provided with your radio is probably just fine for everyday use. If you would like to try a different mic, your radio's operating manual lists the specifications for adding an external microphone. You need to match (approximately) the *impedance* of the microphone and radio (see Chapter 17).

Microphones are extremely complex instruments. They can have any one of a variety of *patterns* that pick up sound from different directions in different ways.

For detailed information about how microphones work, Audio-Technica, Inc. provides a good tutorial on its Web site (www.audio-technica.com/using/mphones/guide/micdoes.html). The online encyclopedia, Wikipedia offers a great entry about mics, as well (www.wikipedia.org/wiki/microphone).

Here's a summary of several types of microphones:

- **Boom:** A microphone mounted on a flexible, extendible arm or tube. Boom mics are placed in front of an announcer or operator working at a desk.

- **Carbon:** A low-impedance microphone not noted for its *fidelity* (in other words, high and low frequencies are lost). Carbon mics have a high output level and are very rugged; they're often used with PA and paging systems.

- **Cardioid:** A mic that rejects sounds from the rear, but receives sounds fairly well in other directions.

✔ **Desk:** A microphone designed to be placed on a desktop. The operator leans toward it when speaking.

✔ **Dynamic:** A medium-impedance microphone that offers good fidelity and a wide range of styles and response patterns. These are the most popular type of microphone.

✔ **Electret or Condensor:** A high-impedance microphone that requires a dc voltage supplied from the radio's mic connector to operate. Miniature electrets are common in hand mics and are built in to recording equipment. Larger electrets can provide exceptional fidelity and are often used in professional recording.

✔ **Hand:** A microphone designed to be held in the palm of the hand.

✔ **Headset:** A microphone/headphone combination, usually including a miniature boom mic.

✔ **Noise-Canceling:** A type of mic that rejects all sounds except those coming from a source very close to the microphone, such as the speaker's mouth.

✔ **Omnidirectional:** A mic that receives sound equally well from all directions.

✔ **Unidirectional:** A mic that receives sound best from one direction — right in front of the microphone.

If that new microphone you just bought doesn't work with your radio or provides muffled, weak audio, perhaps you have an *electret mismatch.* An electret microphone must have a dc voltage applied to work properly. If the voltage is not present, there is no, or very weak, output. Conversely, applying dc voltage to a dynamic microphone restricts its response to sound, resulting in very poor sound quality at a low level. A radio designed for use with an electret microphone won't work with other types. An electret microphone must be used with a radio that provides the proper voltage.

Adding a computer to your radio operation

Computers are more and more a part of radio every day, as you find described in Chapter 20. Whether you use one for decoding received signals or for keeping track of who's on what frequency, a computer may be one of the first additions you make to your radio shack. When it comes to installation, your main concerns should be comfort and safety. You want to make sure that you can comfortably use the computer and that all the necessary cabling is safe and out of the way. Figures 18-3 and 18-4 give you some idea about how to arrange the computer and radios.

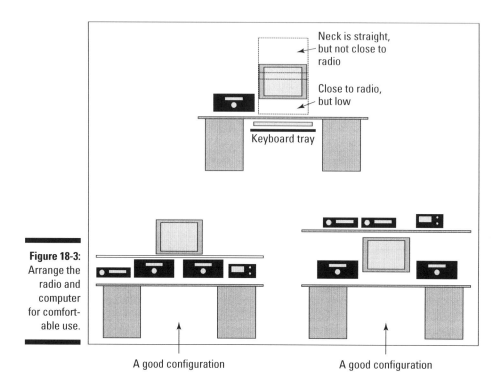

Neck is straight, but not close to radio

Close to radio, but low

Keyboard tray

Figure 18-3:
Arrange the radio and computer for comfortable use.

A good configuration   A good configuration

When connecting a computer to your radio, use good quality data and signal cables. Try to power all the equipment from the same power strip or outlet, using a common safety ground to prevent hum and noise pickup. If you have installed a separate ground system, grounding the computer to that system is a good idea.

Computers often act as low-power radio transmitters, leaking radio signals that are byproducts of the high-speed digital signals inside. Any cable connected to a computer can act as a small antenna for these signals. You may have noticed collars or sleeves on the end of some of your computer accessory cables. These are *ferrite cores* or *beads* that ambush the cable's antenna aspirations. See Chapter 21 for more discussion on preventing interference from computers and other digital appliances.

Finding the right furniture

The fellow in Figure 18-4 is happy because he is comfortable while operating his radio. Here are a few tips on furniture selection to make your home installation easier and more enjoyable to use:

✔ **The Chair:** Radio hobbyists can spend an incredible number of hours perched in front of a radio. Do your rear end and back a favor by purchasing a well-padded chair with good back support. You'll soon agree that it's a wonderful investment.

✔ **Desks or tables:** The surface on which the radio sits should be deep enough to support your forearms as you tune or operate the radio. You need at least 12 inches to support your arms. A radio that's too high above the surface will quickly make your elbow exquisitely sore as you rest your arm on it to reach the controls.

✔ **Shelves:** If you have several radios or will be using a computer with your radios, you should look for a set of shelves. Computer workstations and other office equipment are often quite useful in the radio room, especially those with adjustable heights. Because most stores don't sell furniture specifically created for radio use (I wish!), you may find some interesting and inexpensive items in other departments. For example, I found a particleboard shoe stacker to be perfect for one small desktop installation!

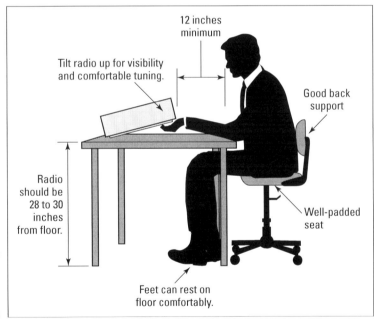

Figure 18-4:
Some of the key elements of making your installation comfortable to use for extended periods of time.

Labels within figure:
12 inches minimum

Tilt radio up for visibility and comfortable tuning.

Good back support

Radio should be 28 to 30 inches from floor.

Well-padded seat

Feet can rest on floor comfortably.

Using antennas and feedlines

What a difference a good antenna makes, even to the least sophisticated radio. You can get no greater bang for the buck in radio than that from a good

Getting ergonomic

Whether you have one radio or ten, being able to get to the back of the operating or listening position is a Good Thing. I have a lot of my equipment on rolling cabinets or carts so that I can quickly swivel them to connect a new cable or do some troubleshooting. Lying on your back or having to crouch and reach gets old fast, making simple tasks difficult.

For more information on *ergonomics* — the study of workplace design — the Cornell University Web site `ergo.human.cornell.edu/ergoguide.html` presents very readable information in the context of computer use. Substitute *radio* for *computer* and you'll get a lot out of a visit to this site.

antenna! Unless your radio's antenna is permanently attached, as with the FRS/GMRS handheld radios of Chapter 4, you have quite a number of antenna options that fit nearly every circumstance. Figure 18-5 shows a few common antenna installations.

Because the different types of antennas used for each service — wire, vertical, and beams — are discussed along with their radios, I treat them generically here. If you're interested in finding the technical details of antennas and browsing the many different types available, check out some of the technical references listed at the end of Chapter 16.

Figure 18-5: These are just a few of the different types of outside antenna installations used for radio.

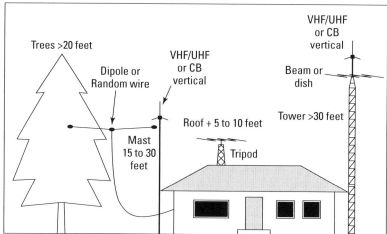

Installation of external antennas may be subject to rental or lease agreements. Homeowner association (HOA) rules may apply to external antennas or restrictive covenants (called *covenants, conditions, and restrictions,* or CC & Rs) may also be written into your property deed. It's best to check these before you begin purchasing materials. If having an outside antenna is important to you, check the fine print before buying, renting, or leasing. Otherwise, plan on installing the antenna in your attic or using a temporary antenna that can be taken down between uses.

Power lines may not present as dramatic a hazard as lightning (see the sidebar, "Being aware of lightning"), but every year more people are injured by them than they are by lightning strikes. When you're installing an antenna outside, check and double-check that you (and the antenna) are well clear of power lines, including the electrical service line from the power pole to your home.

Never run feedlines or wire antennas over or under a power line. When erecting a mast, keep the mast 1½ times its height (including the antenna) from the nearest power line, as shown in Figure 18-6. Not only do masts and antennas fall, but wind can blow them into power lines, as well. Don't join the ranks of the silent operators because of careless antenna installation!

Setting up masts and towers

If you're unable to mount your antenna directly on a building or other convenient support, you have to use a mast or tower. Masts are made of pieces of metal tubing that either nest together telescopically (this structure is referred to as a *push-up mast*) or stack end to end, such as the familiar TV mast sections sold by RadioShack and other companies. You can use masts specifically manufactured to support antennas or use sections of pipe; the choice you make should be based on whichever is sturdy enough to hold your antenna.

Being aware of lightning

Lightning can present quite a hazard in many parts of the world. The lightning frequency map at www.lightningsafety.noaa.gov/lightning_map.htm gives a pretty clear picture of the areas of the U.S. that are home to frequent lightning strikes. Wherever you live, mounting structures should be well grounded and a *feedline lightning arrestor* should be used. These devices install in the feedline and conduct lightning energy to a nearby ground connection or ground rod. If you live in a high-frequency area, I suggest that you follow the advice of a professional installer familiar with the area. For more information on the fascinating physical phenomenon of lightning, check out the links at directory.google.com/Top/Science/Earth_Sciences/Meteorology/Weather_Phenomena/Thunderstorms_and_Lightning/.

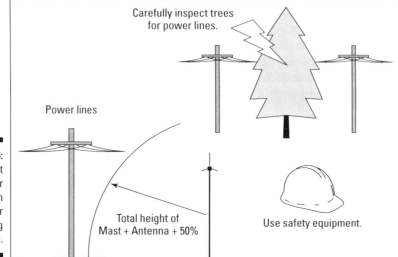

Figure 18-6:
Watch out
for power
lines when
erecting or
installing
antennas.

Carefully inspect trees
for power lines.

Power lines

Total height of
Mast + Antenna + 50%

Use safety equipment.

Larger antennas, such as beams for Citizens Band or amateur (ham) radio, require something sturdier than a mast. Heights of more than 30 or 40 feet are also beyond the ability of ordinary masts. For these jobs, you need to put up a tower. Towers come in sections of 8 or 10 feet and are erected one section at a time. A tower sits on a concrete base and can be *self-supporting* or *guyed*. *Crank-up* towers have telescoping sections that are raised with a winch. Towers are sturdy enough to support a climber (wearing a safety harness, of course) and are quite safe if installed properly and maintained regularly. Antennas are either mounted on a *stub mast* at the top of the tower or from *sidearms* or *side mounts* that attach to the tower somewhere in the middle. Table 18-1 lists the biggest tower manufacturers.

Roof-mounted towers and *tripod mounts* are good compromises in residential areas if a mast or tower won't work. These options bolt through the roof to the supporting trusses. Small roof-mounted towers can only support a section or two of TV mast, but the larger ones extend 20 feet above the roof and can support very large beam antennas.

Masts and towers extending more than 10 feet above the highest point of support generally require *guying* (cables that provide extra stability and support) unless they're designed to be free standing. Manufacturers provide complete instructions on what materials are required and how to attach guying cables.

Avoid the temptation to skimp on guying, particularly if the mast is near homes or public areas, or if you plan on climbing the tower. A good discussion of tower safety is provided in Chapter 1 of the *ARRL Antenna Book,* 20th Edition

(ARRL), but you should know right now that a tower is *not* a project for beginners. Either use an experienced installer or rely on the services of someone experienced with handling the necessary work.

Most masts do not require building permits, but towers usually do. Your city or county building permit office has the complete rules. Insurance companies generally consider masts and towers to be *attached structures*. Your insurance agent can tell you whether your homeowner's or renter's insurance covers them.

Table 18-1	Tower Manufacturers
Company	*Web Site*
U.S. Towers	Sold through Ham Radio Outlet (`www.hamradio.com`)
Rohn Industries	`www.rohnnet.com`
Trylon	`www.trylon.com`
Heights Tower	`www.heightstowers.com/products.htm`
Universal Aluminum Towers	Sold through distributors

Avoid buying a used tower or mast. If you can't resist the desire to buy one that is used, be aware of the potential safety problems that you face: Unless the mast or tower has been in storage, exposure to the elements can cause corrosion, weakening welds and supporting members. If it was disassembled improperly, the tower might be damaged in subtle ways that are difficult to detect when you examine it in its disassembled state. A tower or mast that has fallen is often warped, cracked, or otherwise unsafe. Have an expert accompany you to evaluate the material before you buy.

Understanding the importance of rotators for directional antennas

To turn directional antennas requires a *rotator* (not a *rotor*, which is what turns). These come in many different sizes and styles for many different sizes of antennas. Figure 18-7 shows the different ways to use a rotator.

The wind blowing on an antenna causes it to twist against its mount. The higher the wind speed and the larger the surface area of the antenna, the larger the twisting force becomes. That is why rotators are rated in square feet or to their ability to hold an antenna against some maximum wind speed, usually 70 mph. For example, the Hy-Gain AR-40 is rated for 3 square feet of antenna at this speed.

The manufacturer of your antenna specifies the antenna's *wind load* as a certain number of square feet. VHF/UHF scanning beams are typically small enough to be handled safely by a light-duty rotator, but larger CB beams have much higher wind loads. Be sure to select a rotator that's capable of handling that load. Rotators are connected by a four to seven-wire *rotator cable* to a *control box* near the operator. Stores that sell antennas also have a selection of rotators properly rated to handle them.

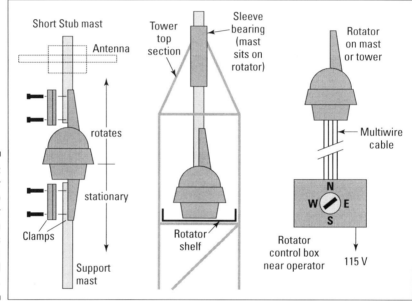

Figure 18-7:
A rotator mounted to the mast or tower both supports and turns a directional antenna.

Choosing a feedline

To make the connection from your radio to the antenna an efficient one requires the proper choice of feedline. Coaxial cable is by far the most common and convenient cable used for feedline because unlike *twin lead* or *ladder line* cables, it can be placed next to metal objects without effect.

All feedlines have *loss* — the dissipation of a signal. The loss becomes greater with frequency and length. Cable materials and construction also affect loss. Table 18-2 lists several popular cables and shows how much signal they lose at different frequencies, assuming the antenna impedance matches that of the feedline. Losses can mount quickly if there is a significant difference. (HF is the high frequency spectrum from 3 to 30 MHz, and UHF covers the ultra high frequency from 300 MHz to 3 GHz. Loss at VHF (very high frequencies) is approximately halfway between HF and UHF loss.) Remember that 3 dB (decibels) represents a loss of half of the signal.

Table 18-2	Relative Cost and Loss of Popular Feedlines			
Type of Coax and Characteristic Impedance	*Outside Diameter (in inches)*	*Cost per Foot Relative to RG-58*	*Loss of 100 Feet at HF in dB*	*Loss of 100 Feet at UHF in dB*
RG-174A (50 ohms)	0.100	80%	6.0 dB	less than 18 dB
RG-58 (50 ohms)	0.195	100%	2.0 dB	11 dB
RG-8X (50 ohms)	0.242	150%	2.0 dB	7 dB
RG-6 (75 ohms)	0.272	80%	1.0 dB	4 dB
RG-213 (50 ohms)	0.405	200%	0.9 dB	5 dB
Twin lead (300 or 450 ohms)	½ inch or 1 inch wide	60% to 130%	0.1 dB	1 dB

As you can see, the thinner cables tend to have higher losses. They're also cheaper per foot. For HF radios, such as CB and shortwave, short runs (50 feet or less) of RG-58 work just fine. For longer runs, use RG-8X or RG-213. As you move to higher–frequency radios, use the lower-loss cables. Why go to all the trouble of putting that antenna way up there if you're going to throw away all the signal in the cable?

Can you use video cable?

In a word, yes. You *can* use video cable. Table 18-2 indicates that RG-6 cable is relatively small, inexpensive, and loses a small amount of signal, and this coax has a different impedance (75 ohms) than most coax cable intended for radio use. You can use it, but you should be aware of a couple of caveats.

The first concern is that because of the different cable impedance (refer to Chapter 17 for a refresher on impedance), 75-ohm cables don't present exactly the right electrical load to the radio, causing a slight increase in loss. Frankly, for ordinary use, the additional loss is minimal and unless you're running high power as an amateur radio operator, you won't notice it.

The second concern is that the UHF connectors often found on radios don't fit RG-6 cable. Neither do the common variants of BNC and phono connectors. Purchase an F-type connector made for the type of RG-6 you use and an adapter to connect it to the radio or antenna.

In short, for all but the most demanding installations (high-performance radio or an antenna situation that requires a long run of cable), RG-6 is a good bargain. Do not use the thin, short cables intended for home entertainment video use, however. Cable quality tends to be poor, connectors are weakly attached, and they are too leaky for radio use.

Cable is drastically cheaper when you buy it in quantity. Purchasing spools of 500 feet or more is the best bargain. Split a spool with friends and share the savings!

Getting the feedline to the radio

I cover selecting and attaching connectors later in this chapter, but for now assume that you have an antenna, a radio, and a spool of cable. (I'm assuming that you're using coaxial cable.) Getting the cable into the house can be as much work as raising the antenna! Although the instructions here are simple, you should do a lot of planning and have a guru around for moral and techno-logical support. Here are some do's and don'ts:

✔ **Do start at the antenna.** Install a suitable connector only on the antenna end of the cable, leaving the other end bare. After the cable is installed; then put on the other connector. It's much easier to install connectors inside than on the roof! Leave yourself plenty of extra cable to avoid cutting it too short.

✔ **Do use pliers.** For threaded connectors (types F, N, and UHF described earlier), tighten the shell with pliers. Don't be ridiculous, just tighten the connector a little past "finger tight."

✔ **Don't let water get into the connection.** Coat the connection with a con-nector sealing putty such as Coax-Seal, a liquid insulating compound, or a tight wrapping of high-quality electrical tape.

I don't recommend silicone compounds because some of them give off corrosive chemicals while curing. Also, silicone compounds are a *total* pain to remove.

✔ **Do leave a *drip loop* where the cable is to enter the building.** The drip loop should be several inches so the cable has to rise to get to the entry point. A drip loop prevents water from running down the cable and up to the wall where it might leak inside. The entry point is where you should install a lightning arrestor such as the Alpha-Delta Transitrap LT (www. alphadeltacom.com) and a ground rod.

✔ **Don't put extra weight on the antenna connection.** If the cable will be hanging from the antenna, tape it securely to the insulator, mast, or boom to take the weight and flexing off of the antenna connection.

✔ **Do make a rotation loop for rotating antennas.** If the antenna rotates, make a rotation loop by twice wrapping the cable loosely around the mast above the rotator. This allows the cable to flex with the rotation. Below the rotator, attach the cable to the mast or tower leg with electri-cal tape or short lengths of wire twisted around the cable and support. Secure the cable to the mast or tower every 10 feet or so to prevent the cable from flapping in the wind.

Direct-burial cable that has a special water-resistant jacket can be installed in a narrow slit in the lawn or in a PVC conduit buried in a shallow trench. If you decide to use conduit, be sure to make it large enough to pass the connector on the end of the cable. If more cables are to be installed later, be sure to leave a pulling wire in the conduit and make the inside diameter large enough for all the cables and a connector.

✔ **Do use nail-down cable clips (don't use cable staples).** Secure the cable to a building with nail-down cable clips (I prefer RadioShack part number 278-1661) or screw in insulated standoffs. Cable staples, especially the kind intended for use with electrical power wiring, crimp the feedline, often damaging it beyond repair.

✔ **Don't pinch coax or bend it sharply.** Coax needs to be kept round and all bends must be gentle.

✔ **Do drill a big enough hole.** If you drill a hole through the wall, remember that you will either have to make a big enough hole for the connector or cut off the connector later! RadioShack and hardware stores have the necessary supplies for running the cable through a wall and inside your home. Pack the hole with stainless steel wool or fiberglass insulation to prevent unwanted insect and animal visitors.

✔ **Do use a window if you can't drill a hole in the wall.** You can remove one pane of a multipane window and replace it with a clear plastic panel. Then you can drill all the holes you want or use bulkhead connectors to get your signals inside. Depending on your security concerns, you can also lift the window and insert a panel through which the cables can be run (see Chapter 17 for more on connectors). You can restore security by purchasing a window lock.

Getting Your Mobile Installation Rolling

You have all the same equipment considerations in an automobile as at home, plus new issues like vibration, temperature, and security to worry about. Just as with a radio setup at home, if you think things through first and take your time, you'll get the job done right without wasted effort. The key is to be informed! Here are the places you should go first for information:

✔ **The manufacturer of your car:** Ask the dealer for service bulletins or guides for installing mobile radios. Because they sell so many cars and trucks to fleet owners, General Motors, DaimlerChrysler, and Ford each have a Web site discussing how to install mobile radios in their cars and trucks. These sites provide excellent reading, whether you have one of their cars or not.

- General Motors: `service.gm.com/techlineinfo/radio.html`
- DaimlerChrysler: `www.arrl.org/tis/info/pdf/INSTG01.pdf`
- Ford: `www.fordemc.com/docs/download/Mobile_Radio_Guide.pdf`

✔ **General Internet search:** Enter your vehicle model and the search term *radio installation* into an Internet search engine such as Google. For example, the first page of hits when I searched for *Contour radio installation* contained `www.gmrsweb.com/gmrsbille.html`. This site shows how to install mobile radios in a car very much like my own! And that's just the beginning. You can find many more sites discussing how to install audio equipment in cars. Because the issues of obtaining power, routing cables, and mounting equipment are very similar, these are good sites to visit. With some careful searching, you can turn up some information about your vehicle. You may even be able to make use of some of the installation tools and materials available from audio installer sites, such as `www.installer.com`.

✔ **Clubs and organizations whose members use mobile radios:** Look for off-road vehicle clubs, four-wheeling groups, scanner clubs, CB clubs, RV organizations, and so on. For example, the Radio Reference Web site (`www.radioreference.com`) is one of the largest organizations for scanner and CB users. Ham radio club Web sites are also good sources of information because, from a mechanical and electrical point of view, the installation process is similar.

Understanding vehicle radio safety issues

There are three aspects of mobile radio safety to consider: mechanical, electrical, and driving. All are important to your successful radio operating on the road.

Installing marine electronics

For those of you with boats, marine electronics installation is very similar to auto installation, with one exception — there's a lot more water! Seriously, the marine environment is not kind to electrical and electronic systems. Before installing your marine radio, I suggest that you browse this section; then proceed to a reference that deals specifically with marine electronics, such as:

✔ *The Marine Electrical and Electronics Bible,* by John C. Payne (A & C Black)

✔ *The Powerboater's Guide to Electrical Systems: Maintenance, Troubleshooting, and Improvements,* by Edwin R. Sherman (International Marine/McGraw-Hill)

Both books show you the special techniques and terminology that apply to shipboard electronics.

Mechanical safety means mounting equipment properly

Mechanical safety means attaching your equipment securely, properly placing antennas and accessories so that you minimize the possibility of endangering yourself or your passengers in an accident.

If you plan on using a *mobile rig* intended to be permanently mounted in a vehicle, resist the temptation to stuff the equipment next to your seat or let it sit on a seat or a dashboard. In an accident, the radio equipment will come loose and fly around inside the passenger compartment at high speed. The stuff that's left unsecured inside a car causes a lot of unnecessary injuries. You should properly mount or secure radio equipment, even if the use is just temporary. External antennas should be placed where a pedestrian (or you!) won't be easily poked by the antenna's tip. You'll put your eye out with that thing! (Yes, Mom. . . .)

Electrical safety means maintaining your connections

Electrical safety primarily concerns proper power connections. Resist the urge to throw things together in a hurry to get on the road. I once nearly set a vehicle on fire because I got sloppy with dc power under the dashboard. Don't let this happen to you! Take the time to identify the proper circuits, use automotive electrical hardware, protect your wiring from vibration and accidental shorts, and ***never*** make an unfused connection to your vehicle's battery.

Driving safety means . . . well, driving safely

Don't let your radio installation and operation impair your driving. If you tune or adjust your radio frequently, place the radio or its controls somewhere that minimizes the amount of time your eyes have to leave the road. Set up memory channels to reduce the amount of fiddling around you're required to do. If you find yourself distracted by radio use, pull over to talk. Don't install equipment where it obscures vehicle controls or obstructs their operation. Just don't.

Setting up a power supply for your car radio

For low-power radios, you can obtain dc current from a cigarette lighter socket by using a lighter plug cable or adapter. Late model autos that may not have a lighter socket often have some kind of accessory power socket for powering laptop computers, mobile phones, and so on. These circuits are usually fused at 5 to 10 amps, which can run a CB or mobile radio with up to about 50 watts of output. (Handheld radios take much less current.)

Resist the urge to put in a larger fuse and run a higher-power radio. The wiring for accessory circuits is not rated for the heavier load and may heat up. At the least, the voltage at your radio will drop when transmitting because of the small wire size.

Here are some other important do's and don'ts:

✔ **Do disconnect the battery whenever you work on a vehicle's power system.** Disconnecting the battery prevents an accident because it prevents destructive and dangerous short circuits.

✔ **Do be consistent and use red for all positive power wiring and black for ground.** Being consistent with the colors lets everyone who works on your car know what the wires are used for. Save other color wire for controls and audio connections.

Auto part stores have all the parts you need for properly obtaining power from a car's (or RV's) power system. If you're a boater, you can find similar materials at a marine supply store, along with corrosion prevention compounds.

✔ **Don't use materials designed for indoor service in a vehicle or boat.** Vehicle and marine service is a more severe environment. Vibration, temperature extremes, moisture, and exposure to oils and greases can quickly overwhelm light-duty home wiring materials. Figure 18-8 shows a few examples of sources that are appropriate for a car.

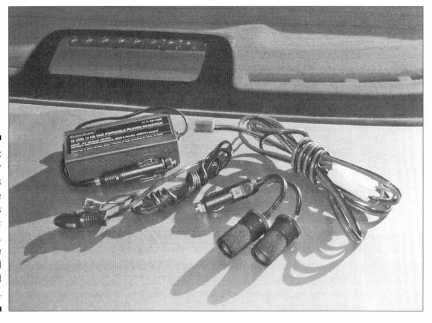

Figure 18-8: Car power sources include lighter plugs and dc to ac inverters. Use in-line fuses and insulated terminals.

Using a fuse tap

If you're working at the vehicle's passenger compartment fuse block, it's often impractical to add a connection to the existing fuse block. Instead, use a *fuse tap* to add a connection to an existing circuit, such as those from Crowbar Electrical Parts (`www.crowbarelectricalparts.com`). Click the Fuse Holders button to find the adapters.

You should always place a fuse tap on the hot side of the fuse to prevent overloading the fuse. You can find the hot side by removing the fuse and then using a voltmeter. The hot terminal is the one that has voltage with the fuse removed. Your new circuit must be fused, so add an in-line fuse if the power cable doesn't include a fuse. This technique can be used for loads of up to 10 or 15 amps. Connect the radio's ground lead directly to a nearby screw or bracket solidly connected to the vehicle's frame.

Some circuits are only energized when the ignition switch is turned on. If you want to use your radio when the engine is turned off, remove the key and then test for voltage at the fuse block.

Setting up and maintaining battery connections

If your radio (or collection of radios) is likely to draw more current than a fuse tap can handle (see "Using a fuse tap," earlier in this chapter), you should obtain power directly from the vehicle's battery. Connecting directly to the battery prevents stress to the vehicle wiring and results in a decrease in the voltage drop to your equipment. Use the power cable provided by your radio's manufacturer. If the power cable isn't long enough to stretch from the battery, extend the cable with more wire of at least the same size as that in the original cable.

Of course, you have to find a hole in the vehicle's *firewall* between the engine and passenger compartments in order to extend the wiring. If you use an existing hole, it will likely have a rubber *grommet* or other protective material to keep the metal from cutting through the insulation. Use that grommet or protective covering for the vehicle wiring. If you have to drill a new hole, be sure to line it with an automotive grommet from the auto parts store. Don't depend on electrical tape over the wires.

In days gone by, it was often recommended that you run your power cable directly to the battery's terminals. For vehicles with the battery under a seat or in the trunk, this is still the right way. A better solution for today's cars is to run the positive wire to the battery + terminal. Run the negative wire to the point at which the battery ground strap is attached to the vehicle, usually on a fender or other nearby sheet metal. Both leads must be fused to protect the radio and other equipment in case of a short circuit! By connecting the negative lead to the vehicle rather than the battery terminal, you protect the radio from being a ground path for other equipment. Figure 18-9 illustrates three common options for getting power to your radio.

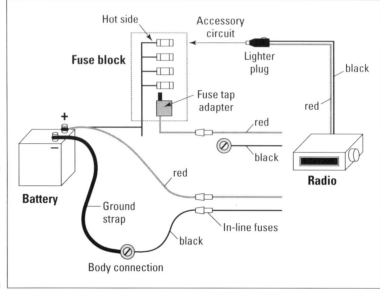

Figure 18-9:
Three ways
to make
proper
power
connec-
tions.

If you're installing a CB radio, think about hooking up your PA speaker at the same time you make your power connections. The wiring can often be run through the same holes, saving you another job later. Here are some additional tips:

✔ As you make your connections, secure the wires to supporting harnesses or brackets with nylon wire ties or electrical tape rated for automotive use. (At high and low temperatures, ordinary electrical tape adhesive will lose its grip.)

✔ Keep the wires away from moving parts and hot assemblies (such as exhaust system components or heater ducts).

✔ Protect the wires from sharp metal edges that might cut through the insulation aided by the vehicle's vibration.

✔ Don't position a connector or terminal so that the wire is pulling on it — over time, vibration will work the wire loose.

Finding a home for the radio in your car

In today's crowded cars, finding room for a radio can be a challenge. How do you find a place for the radio so that it's where you can see it without it obstructing your ability to drive or getting in the way of your passengers? (If you haven't yet bought a radio, consider where you'd like to install it before you start shopping.) Browse the Internet looking for examples of radios mounted in your car or a similar model. Check with your dealer or a professional installer for ideas.

In general, you want to find a location from which the front panel is easy to see and all controls are reachable without fumbling around. The radio must be mounted securely and have sufficient airflow to stay cool. In order of desirability, here are several options:

- ✔ **Inside the dashboard:** This location is the best, but is usually filled with other electronics and controls. If your radio has a detachable front panel, you can usually find a home for it here. A drawback is that this is often a very difficult place to add equipment.

- ✔ **In the forward center console:** Look for blind panels that cover unused accessory mounting bays. You can often remove ashtrays and storage compartments, as well. Not as difficult as the dashboard, but still crowded.

- ✔ **Under the dashboard:** This location can be hard to see (especially when you're driving), but it's an easy spot for installation purposes. Watch out for dangling mic cords or knee-gouging corners (especially if you have a manual transmission — ouch!). The radio is difficult to operate if you're not in the correct seat.

- ✔ **On or to the side of the drive shaft hump:** This is also an easy installation location, but often invades the legroom space and makes viewing and operating inconvenient.

- ✔ **On top of the dashboard:** Cosmetically, this is a terrible location. On the other hand, the radio is easy to read and reach. Another downside is that this location exposes the radio to direct sun and can attract unwanted attention from thieves.

- ✔ **Next to a seat:** This is a position of last resort because the seats muffle the audio output and the location of the controls requires taking your eyes off the road for long periods of time. Restricted airflow can also lead to overheating.

You should consider antitheft security. If you intend to remove your radio when you're not in the vehicle, don't mount it where it will be difficult or time consuming to remove. Many manufacturers offer *slide mounts* so that you can quickly get the radio out. Removing the radio should take no more than 10 or 15 seconds. If you can't remove the radio, you may want to try for an installation where the radio can be hidden with a blank panel.

If your radio has a detachable front panel or offers a *remote control head*, your life just got a lot easier. (Be sure to obtain the *separation cable* that connects the panel to the radio if it's a separate option. You may also need an extension for the microphone cable.) The radio can be mounted under a seat or in the trunk, while the only equipment in view is the panel, which is about ¾-inch thick and can be mounted with Velcro strips, as shown in Figure 18-10, or by using a mounting cradle from the manufacturer.

Figure 18-10:
This radio control head is mounted with Velcro strips and the connecting cable is tucked into seams in the dashboard panels.

If the radio is mounted separately, you can use the mobile mounting bracket that came with the radio. This scenario works well for mounting the radio in the trunk. Under the front seat, you can use nylon webbing to secure the radio to one of the seat rails or mounts.

I highly recommend that you place the radio on some foam to protect it from the bumps and bounces of the vehicle. It's also a good idea to protect the radio with a board or panel to prevent back-seat passengers from accidentally kicking the radio. Take care to route all cables so that they aren't pinched by the seat sliding back and forth. If the radio can't be heard from where it's mounted, a *communications speaker* costs less than $25 and can be mounted somewhere out of the way under the dashboard.

A unique puzzle comes along when you try to use a handheld radio or scanner in the car. The best solution is to provide a sturdy pocket for the radio that hangs from the dashboard or console. Obtain a carrying case that's just a bit larger than the radio and attach it to a convenient location. Cables for the antenna, microphone, and power are run to the pocket, where it's a simple matter to plug them into the handheld radio. A fabric or Velcro strap can secure the radio while you're driving. A communications speaker also helps you to hear over road noise.

Using antennas and feedlines in the car

The biggest decision you must make regarding installing a mobile antenna is where to put the darn thing. Nearly all antennas designed for vehicle use are verticals, relying on the metallic body of the vehicle to provide a good ground plane. Consequently, most places around the car where the antenna is in the clear work reasonably well for mounting an antenna. If you place the antenna so that it rests in the middle of a large area of metal, such as the roof or trunk, it will work well.

Most mobile antennas have two parts, the antenna and a *mount.* Here are your options:

- ✔ Permanent mounts are attached to the vehicle with screws or in a hole cut in a body panel.

- ✔ Semipermanent mounts clamp on to some part of the vehicle (they're called *trunk mounts* or *lip mounts*).

- ✔ Temporary mounts attach with easily detachable clips (called *gutter mounts* or *mirror mounts*) or with magnets (called *mag mounts*).

Whichever mount you choose, the antenna is attached to the mount with a threaded coupling of some sort. There are three popular types: NMO, SO-239, and ⅜-inch-24 threaded. Figure 18-11 shows examples of antenna mounts and couplings. Threaded mounts make removing the antenna easy. That way you can prevent theft and take your car to the carwash.

If you frequently disconnect the antenna on a handheld radio, reduce the wear on the antenna connector by placing an adaptor on it. Disconnect at the adaptor and you will avoid wearing out the hard-to-replace connector on the radio.

An *on-glass* antenna attaches directly to the vehicle's glass windows or windshield. The radio signal is passed through the glass from a patch attached on the other side of the glass. Adhesives are used to secure the antenna and the patch inside. These antennas work well, but are not removable. Because of limits to the adhesive's holding strength, on-glass antennas are limited in size to VHF and higher frequencies.

After you've decided where to install the antenna, you'll have to find a way for the feedline to get to the radio. Many mounts come with cable already attached, so one end is already done. There are three things to consider as you look for the best cable route:

- ✔ **Don't pinch or kink the cable:** When routing the cable around doors and windows, don't let glass or metal bear on the cable. Use a low-pressure point in the door seal. Avoid sharp bends. Either pinches or kinks will eventually cause the cable to fail.

✔ **Don't let water into the vehicle:** Where the cable runs past weather seals, try to orient the cable so that water runs *away* from the entry point. Even better is to breach the seal where it is shielded from water. Cutting the seal so that the cable can pass will work if the cut can be sealed around the cable. This technique unfortunately doesn't work for temporary antennas.

✔ **Protect the cable when it gets inside the vehicle:** You will be running cable behind seats and in areas where sharp edges of metal and plastic may be exposed. These hazards will happily chew through an outer jacket over time as the vehicle vibrates. Running the cable behind trim panels and under floor mats protects and hides the cable.

If the cable does not have a connector at the radio end, you'll have to decide whether to trim the cable to length before installing the connector. I recommend leaving the excess cable coiled up under a seat or in the trunk. That way you have extra cable available if you need to move the radio or attach the cable to a different radio or test equipment.

An old radio truth is that any cable cut to length is too short. If you do decide to trim the cable, make a test installation to be sure you have enough before making the cut.

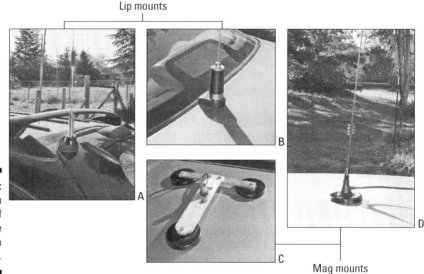

Figure 18-11: Common types of mobile antenna mounts.

Lip mounts

Mag mounts

Choosing and Installing Connectors

Power and radio frequency connectors are found in all radio installations. Chances are that you'll have to install at least a few as you put your radio system together. A poorly installed connector leads to poor performance or failure sooner or later. Why not find out how to do it right?

Crimp terminals

These simple terminals are widely used and are easy to install properly. There are three common types you'll encounter; *ring, spade* or *fork*, and *quick-disconnect*. A *splice* is also available. They have an insulating sleeve over the wire clamp area and the color of the sleeve indicates the sizes of wire that fits the terminals.

- Red: 16 to 22 gauge wire
- Blue: 14 to 16 gauge wire
- Yellow: 10 to 12 gauge wire

All these types of terminals are widely available. Look for them at your local hardware, auto parts, and electronic supply stores.

You'll also need a combination wire stripper/terminal crimping tool that looks like a pair of pliers. (See Chapter 16.) Figure 18-12 shows how to crimp a terminal onto a wire.

Never use a pair of regular pliers to crush the terminal's sleeve — the metal tube doesn't capture the wire properly and it will pull out of the terminal. When you've crimped a terminal, give it a little quality-control yank to be sure you've done a good job. Don't hesitate to cut off a loose crimp and try again. You can't reuse or recrimp them.

RF connectors

Installing an RF connector onto coaxial cable is a craft whose mastery is something to which all radio gurus aspire! If you're just beginning, I suggest that you purchase cable with the connectors already installed; then inspect the connectors so that you know how they are supposed to look. You can find out how to install the connectors by practicing. (I show you the various types of connectors in Chapter 17.)

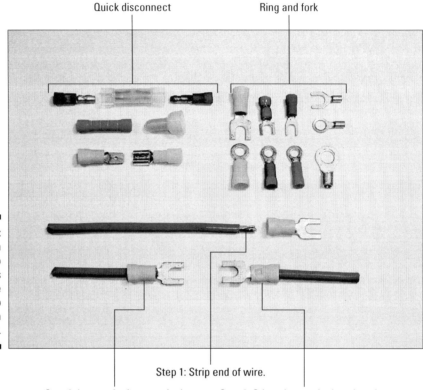

Quick disconnect   Ring and fork

Figure 18-12:
A selection
of crimp
terminals
and the
steps to
making a
good crimp.

Step 1: Strip end of wire.

Step 2: Insert wire into terminal.   Step 3: Crimp the terminal on the wire.

Each type of RF connector — UHF, BNC, N, F, and so on — has its own special installation method. None of them require special tools. You can install some with an ordinary soldering gun, a sharp knife or razor blade, sharp precision wire cutters, and pliers. Others can be crimped on with an inexpensive coaxial cable crimping tool. I also recommend a *bench vise* to help hold the sometimes-stiff cable steady while you work.

The best place for instructions about how to install the connector is, where else, on the connector manufacturer's Web site. I find that the Amphenol Web site has the best information on all sorts of RF connectors, except F connectors. Browse to www.amphenolrf.com and find the connector type you're interested in, such as BNC, UHF, or N, on the left-hand side of the page. Click the appropriate link and look for "Assembly Instructions" on the page that appears. A complete PDF file describing the procedure for each type of connector can be downloaded. If you want an excellent pictorial that shows how to install F connectors, check out `www.interstateelectronics.com/howto/coaxterm.html`.

The American Radio Relay League (ARRL) Technical Information Service's Homebrew page features a nice two-part how-to article on attaching all sorts of connectors commonly encountered by radio enthusiasts. Browse to `www.arrl.org/tis/info/homebrew.html` and look for "Connectors for (Almost) All Occasions — Parts 1 and 2." The page also has a number of other useful articles and Web links on radio construction techniques.

Chapter 19

Getting a Charge Out of Batteries

In This Chapter

▸ Choosing between disposable and rechargeable batteries

▸ Knowing how to charge and discharge batteries properly

▸ Following battery safety precautions

*1*f you have a portable radio set, you've got a lot more freedom in your communications. Of course, portability usually requires battery power — or a very, very long power cord. Even solar power aficionados rely on batteries to store excess energy through the day. Batteries are everywhere and in every sort of device, and it seems there's a new type of battery every month! In this chapter, I review the different battery types so you can decide what battery is right for your radio.

Getting Battery Basics

All the jargon associated with batteries makes a whole lot more sense if you know a little about what makes a battery go. In the simplest terms, a *battery* is a device that produces electricity through a chemical reaction. In this age of gadgets, you use several items every day that require batteries — your car, you laptop computer, your cellphone, your digital camera, and, of course, your radio. Although you don't have to know chemistry to understand why batteries are important, you should know a little something about batteries because if you use your portable radio frequently, you can end up spending a lot of cash on inefficient batteries.

You can find a much more complete description of battery operation at `science.howstuffworks.com/battery1.htm`.

The standard voltage of the average battery varies between 1.2 to 1.5 volts. The rectangular 9 volt (9 V) battery is really made from six 1.5 V batteries connected end to end so that their voltages add up to 9 V. *Lead-acid* batteries, such as those used in a car or a *gel-cell,* provide 2 V per cell. Putting six of them together end to end makes a 12 V battery, which is standard for a car battery. Batteries made with lithium have a characteristic voltage of between 3 and 3.3 V. These are just a few examples.

Batteries come in different sizes and shapes because they have different jobs to do (see Figure 19-1).

Ah . . . Introducing Amp Hours and Characteristic Voltage

Batteries have two primary ratings:

- ✔ **Characteristic voltage:** The relatively constant voltage between the two battery terminals. This measurement is based entirely upon the chemicals used to transfer electrons. (See the sidebar, elsewhere in this chapter, "How batteries work," for more information on characteristic voltage.)

- ✔ **Energy capacity, measured in *ampere hours* or *Ah:*** Because a battery's characteristic voltage is relatively constant, the only change between a little battery and a big battery of the same chemistry is how long the battery can supply current before it dies or requires recharging. If you multiply ampere hours by the battery voltage, you can calculate watt hours (watt hours are the units of measurement used on most electric bills).

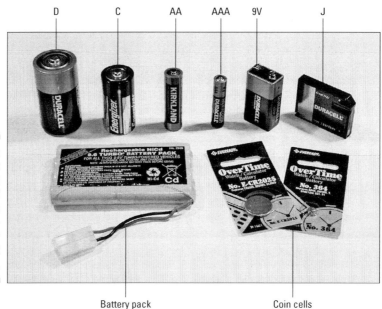

D   C   AA   AAA   9V   J

Figure 19-1:
Different sizes and shapes of batteries are available for different uses.

Battery pack   Coin cells

How batteries work

The chemical reaction can generate electricity because chemicals hold on to their *electrons* (the smallest particle of electric charge) with different strengths. The strength is called *electropotential*. When two chemicals that have different electropotentials are brought into contact, electrons transfer to the chemical that wants the electrons the most. You can't just mix the chemicals together; if you do, the electrons move without going through a *circuit*. And, of course, you need a circuit if you want to make the electricity do some useful work, like run a radio. A circuit is an electrical path along which electricity flows.

If two chemicals are kept from mixing because they are separated by some kind of barrier, they quickly build up an imbalance of electrons as the greedy chemical collects them. This imbalance of electrons keeps the remaining electrons on the other side of the barrier. The resulting imbalance between one side and the other creates the voltage difference you measure between the terminals of the battery. Every battery has two terminals, positive (+), and negative (-). If you connect a circuit from

one terminal to the other, the voltage pushes the electrons through the circuit, supplying energy to run the electronics.

The voltage between the terminals is called the *characteristic voltage* of the battery. The characteristic voltage depends on the type of chemicals used. When you go to the store and see alkaline, nickel cadmium (NiCd), or lithium-ion (Li-Ion) batteries, they all have a slightly different voltage. Battery manufacturers have come up with a number of chemical combinations that produce voltage.

The terms *dry-cell* and *wet-cell* battery refer to the makeup of the *electrolyte*, the barrier that separates the chemicals. Most types of batteries are a little bit wet, or at least damp. A dry-cell battery's electrolyte is a soggy paste, for example. A wet-cell battery, such as the one in your car, actually has a completely liquid sulfuric acid electrolyte (danger!). Gel-cells have the same chemicals in them as your car battery, but instead of liquid sulfuric acid, the electrolyte has been made into a gel to prevent spilling.

If a battery can supply one ampere of current for an hour before exhausting its chemicals, it's rated at 1 Ah. A rechargeable NiCd battery with a rating of 1200 mAh can supply 1200 mA (or 1.2 A) for an hour. The actual rating is more complicated than that, but you get the idea. Table 19-1 shows a number of common battery sizes, chemistries, and energy ratings.

Table 19-1	Common Battery Types and Ratings for Electronics		
Battery Style	**Chemistry**	**Voltage (With a Full Charge)**	**Energy Rating (Average)**
AAA	Alkaline (disposable)	1.5 V	1100 mAh
AA	Alkaline (disposable)	1.5 V	2600-3200 mAh

(continued)

Table 19-1 *(continued)*

Battery Style	Chemistry	Voltage (With a Full Charge)	Energy Rating (Average)
AA	Zinc-carbon (disposable)	1.5 V	600 mAh
AA	Nickel cadmium (NiCd) (rechargeable)	1.2 V	700 mA-hr
AA	Nickel metal hydride (NiMH) (rechargeable)	1.2 V	1500 to 2200 mAh
AA	Lithium-ion (Li-Ion) (rechargeable)	3.3 – 3.6 V	2100 to 2400 mAh
C	Alkaline (disposable)	1.5 V	7500 mAh
D	Alkaline (disposable)	1.5 V	14000 mAh
9V	Alkaline (disposable)	1.5 V	580 mAh
9V	Nickel cadmium (NiCd) (rechargeable)	9 V	110 mAh
9V	Nickel metal hydride (rechargeable)	9 V	150 mAh
Coin Cells	Lithium-ion (Li-Ion) (rechargeable)	3 to 3.3 V	25 to 1000 mAh

Disposable Batteries versus Rechargeable Batteries

With some kinds of batteries, when the chemicals have transferred all of their electrons, that's it, and the battery is dead. Those are disposable or *non-rechargeable* batteries. Other batteries are *rechargeable,* which means the chemical reaction can be run in both directions; one way creates electricity (called *discharging*) and the other way stores energy (called *charging*).

Don't try to recharge a disposable battery, because the chemicals just won't change back to the way they were. Attempting to recharge a disposable battery runs the risk of heating up the battery or causing major corrosion.

Disposable batteries and rechargeable batteries each have their own merits, which I discuss later in this chapter. If you're trying to decide between the various battery types, try regular, old-fashioned disposables and see how long they last. If you find that they die within a few days or weeks, you should really consider the more expensive rechargeable options.

Eventually, all batteries must be disposed of. Even rechargeable batteries eventually die. Be sure to keep extras around even if you rely primarily on rechargeable, long-life batteries.

 One factor in evaluating the quality of any battery (rechargeable and disposable alike) is the battery's *discharge curve*. In Figure 19-2, you can see a graphical representation of how battery voltage changes as its energy is used up. The two performance leaders are alkaline batteries (which are disposable) and NiMH batteries (which are rechargeable).

The perfect battery would provide a constant voltage (represented in the figure as a horizontal line) until it is completely exhausted, and then drop vertically to zero voltage. Real batteries try their best, but can't quite meet that standard. As they are discharged, their output voltage gradually drops to about 80 to 90 percent of the full-charge voltage. At some point, their output voltage starts to drop rapidly. If you think of the discharge curve as representing a battery-powered flashlight's brightness, you will get an idea of how the battery performs as it is discharged. The following sections rate all battery types, categorizing performance and cost effectiveness.

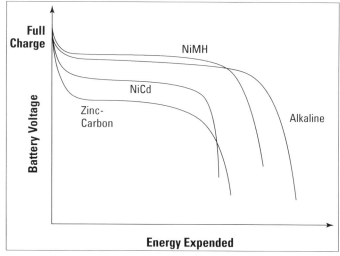

Figure 19-2:
Discharge curves for several types of batteries.

Disposable batteries

You're probably most familiar with disposable batteries; after all, Energizer bunnies and coppertops have immortalized the alkaline battery since the 1960s. Here are ratings for the two most common disposables around.

Disposable batteries contain chemicals that aren't exactly poisonous, but that aren't good for the ground or water supply. Instead of throwing dead batteries away, ask your local recycling center, electronics shop, or hardware store whether they can be recycled (see "Safely disposing of batteries," later in this chapter.)

Zinc-carbon

I don't recommend zinc-carbon batteries for your radios. Although they function fine for a while, they have relatively short lives. Also, between the rod and case of each zinc-carbon battery is a paste saturated with a weak acid — the very same acid that leaks out and ruins your flashlight when you leave discharged batteries inside. All in all, this is not a winning combination.

If a battery's voltage drops too quickly, many radios turn off before the battery's energy is completely used. If your radio suddenly stops functioning, the first thing you should do is check the batteries.

Alkaline

The alkaline battery is a disposable battery that demonstrates an improvement in chemistry over zinc-carbon batteries. It uses an alkaline solution that is much less corrosive than the weak acid used in zinc-carbon batteries. Alkaline batteries also have a much higher *energy density* than zinc-carbon batteries, meaning that more energy is packed into the same volume. You can see evidence of this by reviewing the higher Ah ratings for alkaline batteries in Table 19-1.

The alkaline battery's discharge curve is much flatter than that of the zinc-carbon battery. The output voltage stays steady until much more of the stored energy is discharged. Because it contains more energy to begin with, the alkaline battery runs electronic devices (like your radio) for a much longer time than the zinc-carbon battery.

Strengths

 ✔ Widely available with a high energy rating

 ✔ Does not corrode

Weaknesses

 ✔ Expensive when used continuously

 ✔ Not rechargeable

 ✔ Heavier than alternatives

Rechargeable batteries

Rechargeable batteries generally give you the best of all possible worlds. They have long lives and are also rechargeable. Of course, not all rechargeable batteries were created equal, and cost is a consideration. The following sections sort through the issues and benefits of these batteries.

Nickel cadmium (NiCd)

The nickel cadmium battery (NiCd) was the first popular rechargeable battery type and is found in all sorts of rechargeable appliances, such as toothbrushes, drills, cordless and mobile phones, and model cars. Nifty as they are, there are better alternatives.

Cadmium is a toxic metal; although individual batteries pose no danger when used in electronic devices, after the battery finally dies, you're faced with disposal problems in landfills. Cadmium can be leached into the ground soil, and this is not a good thing. (See the section, later in this chapter, called "Safely disposing of batteries" for more information about proper battery disposal.)

NiCd batteries (pronounced "NYE-cad") have a decent energy rating, but they don't last as long as alkaline batteries. Their discharge curve is a little weak towards the end of charge, but they can be recharged many times at little expense.

Early generations of NiCds had a *memory effect* that gradually reduced the battery's energy rating when recharged unless special conditioning was performed. The NiCd batteries sold today are largely free of that problem. For low-power radios, NiCd batteries are fine, but don't get too used to NiCds; they're being phased out in favor of NiMH batteries.

Strengths

- Medium energy rating
- Low cost
- Rechargeable

Weaknesses

- Slight memory effect over many charge/discharge cycles is still possible, although this problem has been largely fixed
- Cadmium is toxic when disposed of improperly
- Lowest energy density of rechargeable battery types

Nickel metal hydride (NiMH)

Nickel metal hydride (NiMH) batteries were developed to better meet the higher power delivery demands of modern electronic gadgets like digital cameras, portable computers, and audio players. The cadmium used in NiCd batteries is replaced with a compound of hydrogen and other metals — voilà, no more toxic metal. Additionally, overall battery performance is greatly improved.

NiMH is the first rechargeable battery type to exceed the alkaline's energy density. As a result, these batteries are rapidly displacing NiCds. NiMH batteries hold their output voltage nearly constant for much longer than the NiCd and are comparable to the performance of top-quality alkaline batteries.

NiMH battery chargers have become quite inexpensive and can recharge battery cells in just a couple of hours; I recommend this battery for your consideration. They'll pay for themselves in no time!

Strengths

✔ Excellent energy density

✔ Can be recharged quickly and for many cycles without memory effects

✔ Less toxic waste to dispose of

Weakness

✔ More expensive than alkalines

Lithium-ion (Li-Ion)

The next step in battery development, Li-Ion batteries can be recharged many times over. These cells are often combined into battery packs and used to power laptop computers. They are just becoming available as separate cells.

Strengths

✔ Excellent energy density

✔ Can be recharged quickly and for many cycles

Weakness

✔ Most expensive of rechargeable types

Lead-acid

If you need a lot of energy, only a lead-acid battery will do the job. Although the lead compounds and strong acid electrolyte pose some toxicity problems, these batteries are inexpensive to manufacture and easy to recycle. You can see why this has been the battery of choice for most high-power uses for nearly 100 years.

Luckily for radio users, the hazards of the acidic electrolyte have been reduced with the creation of the gel-cell battery. Instead of a liquid, the acid is contained in a gel that prevents spillage with only a small loss in energy rating. The gel-cell is a safe and effective means of providing long-term energy storage.

Strengths

✔ Good energy density and inexpensive

✔ Can be recharged many times

Weaknesses

✔ Heavy

✔ Toxic materials and dangerous electrolyte require proper disposal

Exploring the World of Battery Packs

Many handheld radios come with or accept an assembly containing several individual batteries that are permanently wired together. These *battery packs* can be made to fit almost any shape required and in a variety of voltages.

If you crack open the case of a dead battery pack, you see several cells connected end to end in a series with welded tabs. It's generally not practical for the hobbyist to rebuild battery packs, but they can often be traded in or recycled.

Some two-way radios are designed to accept battery packs of several different voltages, varying their output power as the voltage changes. You can choose a lightweight pack with a low energy rating or a heavier pack that gives more transmitter output. For my own ham radios I have one or two of each type of battery pack.

Many radios allow you to choose to use either individual batteries or a rechargeable pack. It's hard to beat the convenience of using a drop-in charger by using one or two battery packs instead. However, if you intend to use the radios for emergency communications, it's wise to have a battery pack that accepts individual batteries, usually AA cells. After all, when the power is off, you can't use your charger. At those times, being able to drop in a set of alkaline batteries and keep going is priceless!

Following Basic Battery Tips

Here, in no particular order, are some good ideas for getting the most out of your batteries:

✔ **Buy disposable batteries in bulk.** Buying alkaline batteries four at a time at the grocery store is convenient (and may seem less expensive at the point of purchase), but in the long run, it's incredibly expensive. Warehouse stores and online suppliers often have bulk packs of 50 or more batteries at a fraction of the price per battery if you were to buy them in small quantities.

✔ **Refrigerate spare batteries.** Because a battery is a chemical device and chemistry runs slower at lower temperatures, you'll prolong the life of your batteries by keeping them cold. Don't freeze them! The water in the electrolyte expands when frozen; battery cases may crack, ruining the batteries.

✔ **Have spares always.** When buying batteries, always have extras for emergencies and periods of prolonged use. Don't be caught discharged!

Because batteries allow you to take your radio on the road, you should be extra diligent about bringing spares with you. Who knows when and where you'll find a plug for your charger?

✔ **Group rechargeable batteries in sets.** A set of batteries is only as strong as its weakest cell, so try to keep sets together as the batteries age. Battery suppliers sell inexpensive plastic cases that hold a few batteries. Don't let one weak cell spoil the whole set.

✔ **Regularly condition rechargeable batteries to help keep them in top shape.** If you have several sets of batteries or battery packs, rotate them through a charge/discharge cycle on your charger every few months.

Adhering to the Rules of Battery Safety

Used properly, a battery is a safe and effective way of storing electrical energy. However, a battery stores a *lot* of energy and uses *concentrated chemicals* to do so. Think about that.

Letting the energy out too quickly can wreck the battery — and wreck whatever it's connected to. Exposing the chemicals or putting them under too much stress can result in danger to you and damage to the battery.

Charging and discharging batteries safely

To get the longest life and best performance out of your investment in rechargeable batteries, you should use a charger designed specifically for that battery type. For example, a NiCd battery charger won't properly charge NiMH batteries. An improper charger may even damage the batteries or become damaged itself.

Good advice in general (not just about batteries) is to do what the manufacturer tells you to do. The manufacturer wants you to have good results with its products and to be safe when using them.

Don't try to speed up battery charging with a high-power charger if the batteries aren't rated for it. Limit discharge current to within the battery's ratings. Use chargers on the batteries they're intended for. The reward for pushing the envelope is rarely worth the risks.

Here are some other important things to remember:

- ✔ **Set your smart battery charger to the correct setting.** Smart chargers have a specific method of charging, called a *charging algorithm*. They are able to sense the battery's charge level and adjust the rate of charge so that charge time and stress on the battery are minimized. The optimum algorithm varies with battery chemistry so that NiCd and NiMH batteries should be charged differently. If you have a smart charger, be sure it's set for the right battery type.

- ✔ **Promptly remove charged batteries from your old-style battery charger.** Old-style battery chargers just apply a certain amount of current to the battery until you remove it. Leaving a battery in such a charger can overheat and ruin a battery. Take care that batteries are removed promptly when charged.

- ✔ **Follow the battery manufacturer's guidelines for charging and obtain a proper charger if necessary.** A lead-acid battery charger must be able to switch to *trickle charge* or *float charge* automatically when full charge is reached. This keeps the battery at full charge without overcharging it.

- ✔ **Let your batteries run down every so often.** It does not damage NiCd and NiMH batteries to be completely discharged. In fact, regular *100 percent discharges* help to restore the battery chemistry. Good chargers first discharge these batteries before charging them back up.

- ✔ **Treat your lead-acid batteries right.** *Deep-cycle lead-acid batteries,* such as those intended for RV and marine use, are also able to withstand deep discharges. Regular automobile batteries, however, are not made for that type of use and will be damaged. Only completely discharge batteries that can handle it.

The truly "charged up" reader will enjoy the Web page published by the Battery Tender Company at `batterytender.com/battery_basics.php`. This site offers all you ever wanted to know about how to charge lead-acid batteries.

- ✔ **Don't stress out your batteries.** Batteries can also be stressed by discharging them too rapidly. Short circuits and excessive loads can wreck internal connections and cause electrolytes to boil or vaporize. This type of use calls for special batteries with adequate current surge or pulse ratings.

Storing and handling batteries with care

Here's a short list of do's and don'ts to follow when you're handling radio batteries:

✔ Don't freeze a battery. The water in the electrolyte may expand, cracking the case.

✔ Don't expose a battery to excessive temperatures. High temperatures may result in too much pressure inside the case or damage to the chemicals.

✔ Do keep batteries clean, dry, and off wet floors or shelving. Keeping batteries from getting damp and dirty prevents slow discharge across the battery's surface.

✔ Do use compounds that prevent terminal corrosion for any lead-acid battery that you store or use in an exposed location.

✔ Do take special care with wet cells, such as car batteries. These batteries may leak small amounts of acidic electrolyte that can damage unprotected supports.

Batteries pack a real punch, when it comes to the ability to deliver current. Even AA batteries can put out several amperes if it's short-circuited. This amount of energy is enough to make a battery quite hot — hot enough to burn you or melt a small wire. Larger batteries can get hot enough to explode when they are short-circuited. A large lead-acid battery can start a fire or melt a tool when short-circuited. Respect the energy batteries contain. Protect battery terminals against accidental short circuits. For larger batteries, keep terminal protectors installed at all times.

Safely disposing of batteries

The same chemicals that give batteries their great energy storage ability are also fairly toxic or corrosive. Don't throw old batteries in the trash where they wind up corroding in a landfill, leaching chemicals into the water supply — it may actually be illegal in your area! Hardware stores and battery stores will often recycle your old batteries, including alkalines, and often for free or a small fee. Your municipal or county government may also have a recycling program for batteries. Check www.batteryrecycling.com for more ideas.

Chapter 20

Putting Your Computer to Work

*W*hat gadget doesn't have a computer buried inside it these days? Just about every radio has a microprocessor inside, making it a natural to be connected to a PC. Not only can the computer talk to your radio, but it can decipher what the radio hears, too! In this chapter, I discuss how to select a computer, how to get it connected to your radio, and how to get the connection working. I also discuss some of the symptoms of common problems.

Making Sure Your PC and Radio Are Compatible

There is a huge range of computers and radios. How can you be sure they can work together? In the following sections, I assume that you have a specific radio application in mind, such as logging, radio control and programming (see Chapter 10), or signal decoding (see Chapters 8 and 15). You have to consider two types of compatibility: *software* (whether the computer can run the programs that do what you want to do?) and *hardware* (whether you can connect the radio and computer together).

Determining software compatibility

Regardless of the type of software, asking three fundamental questions can usually help you determine whether it will perform suitably on your computer:

✔ Will the software run on the computer's operating system?

✔ Does the computer have enough resources to run it?

✔ If the software connects to the radio, does the software support the model and version of the radio?

Greater than 90 percent of radio software is designed for use with the Windows XP operating system. The remaining 10 percent of software is split between Macintosh, Linux, and older MS-DOS systems.

With the exception of signal decoding programs (discussed later in this chapter), programs associated with radios don't require a lot of processing power or memory. If you have a computer made in the past two or three years with an average processor speed (300 MHz or better) and a reasonable amount of memory (64 MB or more), it will probably be just fine. If your PC has several hundred MB of disk space available, it's probably okay to run most programs. If you have any questions, consult information about the software's minimum requirements (you can find a list of minimum requirements on the author's Web site or in the user's manual).

If you won't be connecting the computer to the radio or if you only plan on simple radio control tasks, you can often make do with an older computer model. The local PC recycler is a good place to find computer bargains for programs like this.

Higher computer capabilities are required as you add more — and more complicated — radio tasks. Audio recordings and images quickly eat up hard drive space. If you anticipate storing a lot of data, be sure you can transfer it to CD, DVD, or some other backup device. Otherwise, you may need to increase the hard drive capacity.

It's generally easy to determine whether the software supports your radio by looking up *supported models* in the software documentation or on the vendor's Web site. You should also be aware that the software in the radio's microprocessor, called *firmware*, may affect compatibility. If you already own a radio, check its manual to see whether there is a way to read the firmware version; you can usually find this information alongside reset or clear memory instructions. If you haven't yet bought a radio, check with the software vendor to see whether the latest radio versions are supported.

Determining hardware compatibility

The computer and radio can be *interfaced* (connected together) in several ways: *control, audio, peripheral.* All three of these are shown in Figure 20-1. A radio's *control interface* provides access to information about radio settings such as frequency and mode or accepts data to control the radio by tuning it, changing its receiving filters, or reprogramming its memories. For the computer to be hardware compatible, it must have the matching type of interface.

Your radio's owner's manual specifies the radio's interface type. It may be a USB 1.1 port, USB 2.0 port, or a serial port (also called COM or RS-232 port or a legacy port). A high-speed interface, such as FireWire or Ethernet, is generally not required. For USB interfaces, drivers for your operating system should be available from the radio's manufacturer.

Radios with a *proprietary* interface (a nonstandard interface used by one manufacturer) require special cables and probably require special software on the computer, as well — check the radio's manual.

The computer must also have the right type of *audio ports* if the computer is going to be used to decode signals received by the radio. Audio signals from the radio are connected to the computer through standard *sound card* connections (these are usually labeled as Line In and Headphones). The type of sound card required depends on the decoding software. Received signal audio is usually obtained from the radio's headphones or speaker jack. Transmit signal audio is generally connected to the radio's microphone jack. An external *audio interface* may be used as described later in this chapter.

There are several ways in which a computer can effectively press the mic switch and switch your radio into transmit mode. Some use the audio from the computer's sound card, so there's no additional circuitry required. Others rely on an external peripheral called a *keying interface* controlled by a USB, serial, or parallel port on the computer to control a transistor that does the job.

Finally, your system may require some kind of peripheral device that acts as a translator between the computer and radio. A *data converter* is a typical example of this kind of peripheral; it performs the signal decoding and turns it into digital data that can be received by the computer (I discuss data converters in more detail in "Choosing a Signal Interface," later in this chapter). In this case, you'll need to ensure that the computer has the right type and number of interfaces for the peripheral, too. As with the radio control interface, these are usually USB or serial ports.

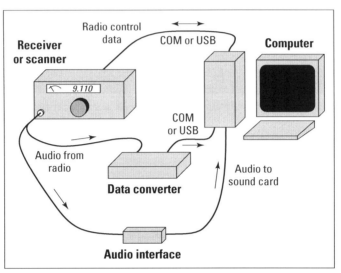

Figure 20-1:
The three types of computer-to-radio interfaces: control, audio, and peripheral. The data converter is shown as a typical type of peripheral device.

Decoding signals

If you are going to use your computer for decoding received signals such as weather fax (Chapter 8) or digital data (Chapter 15), you must make sure that your computer has the right processor speed and the correct sound card's resolution. For example, the popular program for receiving weather fax signals, Mscan Meteo (mscan.com), requires at minimum a 166 MHz Pentium and 24 Mbytes of RAM and a standard (Soundblaster-compatible) sound card. Remember, these are the *minimum* requirements, so if you want to run any other programs on the computer at the same time, I recommend the following minimum resources:

✔ **Processor:** 500 MHz or faster

✔ **RAM:** 64 MB or more

✔ **Sound card:** 16-bit works with most programs

To find out how fast your processor is, choose Start⇨Settings⇨Control Panel⇨ System. On the General tab of the System Properties dialog box that appears you see information about processing speed and RAM. If you don't have your sound card manual handy, you can determine the resolution of the sound card by visiting the Control Panel. Click the Sounds and Audio Devices option. Locate the sound card, copy the manufacturer and model number, and then look it up on the manufacturer's Web site.

Choosing a Signal Interface

You can't just hook your computer and radio directly together. Radios and computers don't always see eye to eye electrically when exchanging small audio signals. If connected directly, the radio often picks up a hum from ac power. This noise (or other noises) can contaminate audio signals, making decoding incoming signals very difficult.

To avoid this problem, you need to put another piece of equipment between your radio and your PC (commonly called an *audio interface* or a *signal interface*). Figure 20-2 shows two different types of audio interfaces, one commercially made, and the other built from a kit (or *homebrewed*).

The hum and noise problem is solved by small transformers that pass the audio signals between the radio to the sound card without a direct connection between them, a type of interface called *isolated* because the two systems (radio and computer) are isolated from each other.

General purpose signal interface

Packet data interface   General purpose signal interface (kit)

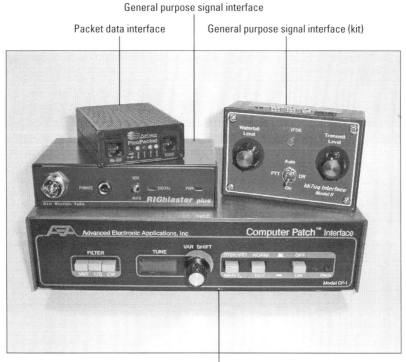

Figure 20-2: Examples of signal interfaces and data converters.

Data converter for radio teletype

Other useful audio interface features include microphone switching (so you can use the radio for voice or digital data), built-in keying interface, audio volume indicators, and a headphone jack. The West Mountain Radio Web site (see the following list) has a comprehensive comparison chart. These vendors make audio interfaces with a variety of features:

- ✔ **Bux Comm RASCAL:** www.packetradio.com
- ✔ **MFJ Enterprises MFJ-1275:** www.mjfenterprises.com
- ✔ **Tigertronics SignaLink:** www.tigertronics.com
- ✔ **West Mountain Radio RIGblaster:** www.westmountainradio.com

If software is available for decoding the signals you want to decode, you just need a signal interface. However, the available software won't always cut it on its own; some signals are too complex for the sound-card-and-computer solution. In these cases a *data converter* (also called a *data controller*) is required. A data converter contains its own audio interface and special electronics to translate received signals into data characters that it then sends to one of the computer's digital interfaces, where the signals are read as text.

The signals used by the SailMail system to send email over HF radio (see Chapter 8) are a good example of data converters in action. This system uses PACTOR III signals (a type of digital transmission) that can only be decoded by an external SCS-PTC data converter made by the SCS company (www.scs-ptc.com). It translates the audio signals received in PACTOR III form to digital characters that can be read by the software.

Data converters are available for a number of digital signals, although sound cards and software are becoming more capable all the time. (See Chapter 15 for more information on digital signals.)

External data converters also allow you to use older (read cheaper) computers to display the text information instead of buying a fancy (read expensive) computer that can handle the software you'd need to do the job. Along with the SCS converter mentioned earlier, here are examples of data converters that can decode multiple types of data signals from your receiver's audio output:

- ✔ **Timewave PK-232:** www.timewave.com
- ✔ **MFJ Enterprises MFJ-1278:** www.mjfenterprises.com
- ✔ **HAL Communications DXP-38:** www.halcomm.com

Making the Connection

After you have the equipment, getting everything hooked up properly is the next hurdle. There are four basic steps:

1. **Control connection:**

   You need to get a computer's USB or COM port connected to the radio.

2. **Radio audio and keying connection:**

   You need to find the radio's audio input and output connections and use the correct cable to connect them to the computer's audio ports, audio interface, or data converter.

3. **Soundcard Connection:**

   You need to connect the radio or signal interface outputs to sound card inputs and outputs.

4. **Level adjustment:**

   You need to get the radio and the sound card level (meaning volume) controls set up properly.

The following sections run you through this process.

Setting up the control connection

The control connection is the simplest of the four steps. USB interfaces only require the use of a cable with the right A- or B-style connector on each end. The USB driver and control or decoding program does the rest.

Serial interfaces require a 9-pin cable with a male or female connector on each end. Computer serial ports COM1 and COM2 use male DB-9 connectors (see Chapter 16). Most serial ports on radios and signal interfaces have female DB-9 connectors. (If you encounter a serial connector with 25 pins, you can probably use a 25-to-9 pin adapter. Check the equipment manual for complete instructions.)

Straight-through connections

The usual configuration for serial interfaces intends for them be connected with a *straight-through* cable. Straight-through cables have pin 1 connected to pin 1 at each end, pin 2 to pin 2, and so on. This should result in data outputs being connected to data inputs. In some cases, often indicated by both pieces of equipment having connectors of the same sex, using a straight-through cable results in data inputs connected together and data outputs connected together. In this case, no data flows and you need a *null-modem adapter* or a *null-modem cable*. These adapters have internal crossovers so that inputs and outputs are connected together again. Having a couple of these adapters in your toolbox is a very handy thing. Search Wikipedia (www.wikipedia.org) entry for the term *RS-232 to* find out all you need to know about this interface.

Setting up the radio audio and keying connection

The radio's audio inputs and outputs are the next hurdle. If you won't be transmitting, you won't be sending any audio to the radio, nor will you be activating or *keying* the transmitter. If you will be sending data, the audio connection or audio interface sends an audio signal in place of the microphone, and a keying interface keys the transmitter as if the PTT switch were pressed.

Audio signals from the radio are obtained through the headphone output, speaker jack, or the microphone connector of some radios. If you are connecting to the computer through an audio interface, the manual has the exact instructions and most manufacturers offer a set of cables along with the interface to help you make the necessary connections. The manufacturer often has proprietary cables that are wired for specific radios.

Connecting the sound card

At this stage, you connect the radio's audio to the computer sound card with three miniature phone jack connections:

- ✔ Line-in (blue)
- ✔ Microphone (pink)
- ✔ Line-out/headphones (green)

Both line connections are stereo connections. The microphone connection may be mono or stereo. If you are not using an audio interface, use the line-in jack. Depending on the audio interface you choose, it uses either the line-in or microphone connection. Most interface manufacturers provide cables. If yours doesn't, you can buy the proper cables at a stereo shop or at RadioShack.

The Web site ePanorama.net (`www.epanorama.net/links/pc_sound.html`) is a good reference if you have questions about connecting to PC sound cards.

Adjusting the levels

When you have your interface wired up, you need to adjust the amplitude of the audio signals or *set the levels* to give the computer or interface a strong enough signal to work with. If you're transmitting, you also have to be sure that the computer can activate the transmitter and also send data at the right level to generate an undistorted or *clean signal.*

Setting the radio's output level is usually just a question of where to set the volume control. The interface or software package manual includes instructions. This is generally not a fussy adjustment. Here are some basic tips:

✔ Listen to a signal directly from the radio. If you can hear it clearly without a lot of static, hiss, or distortion, the sound card is likely to hear it just fine, too.

✔ To set the signal levels, you need to use the audio controls built into your sound card's application software.

Figure 20-3 shows the level control screen for the popular Digipan program (www.digipan.net).

I can't overstress the importance of following the instructions that came with your software or interface. If you follow them, soon you'll be watching characters stream along the screen like something from the movie *The Matrix.*

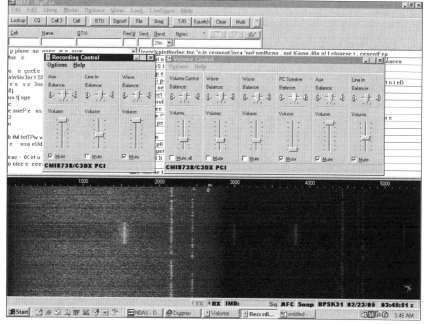

Figure 20-3:
Sound card mixers are used to control audio input and output levels.

Troubleshooting Your Computer and Your Radio

Because I don't know what kind of connection you have between your computer and radio, it would be silly to try and diagnose a problem too specifically. However, I can list the usual suspects and you can figure it out on your own. Follow these basic troubleshooting steps. Continue reading until you find a potential source of your problem:

1. **Confirm that all the power is on.**

   If the power is off on any piece of equipment, you may have solved your problem already. Don't laugh; it happens to all of us!

2. **Check the cables.**

   Are they in the right connectors? Don't be too quick to assume they are. I have spent many an hour chasing a problem that turned out to be a cable plugged into the wrong connector. Are they fully inserted?

   If your problem was a cable or plug issue, fix the problem and repeat the task you originally set out to perform. If not, continue reading the following sections.

When I use the term *radio* in the following sections, I am referring to the radio itself, as well as the signal interface.

Common radio problem #1

The software doesn't know the radio control interface is connected. You can diagnose this problem by checking the computer ports:

✔ **If the radio is connected through a USB port:** Try unplugging the cable from the computer and plugging it back in a few seconds later. If the radio is connected, the computer should detect it and display some kind of message that the radio is present. If the radio is not detected at all, the computer interface may be disabled. If the radio is detected, but unknown, you need to install a driver program from the radio manufacturer.

✔ **If the radio is connected through a serial port:** Check the configuration of the radio and of the computer's serial port. Be sure the computer software is configured to use the correct port. Make sure the specs match. For example, the computer port's data configuration for baud rate, amount of data, start, and stop bits should match your radio's specifications. Some radios also have an address that must be set properly in the computer software. If the radio seems completely dead, try adding a null-modem adapter. (See the sidebar, "Straight-through connections.")

You can fix this problem by finding out whether signals are being received by the computer software:

1. **Listen to the radio directly.**

   Be sure there's something there to be heard in the first place. If you are hearing normal receiver noise or signals, check to be sure the cables to the computer are connected to the right jacks and fully inserted.

2. **Check the computer software's audio input level controls.**

   If you have the interface set up to use a specific mixer input, be sure that input is turned up so the computer can hear the signal.

3. **Make sure you don't have the Mute option checked in the level control screen.**

   Refer to Figure 20-3.

Common radio problem #2

The control data to or from the radio is garbled. This is usually a sign of a configuration problem. Go back through the setup instructions for the software and be sure all settings are okay, such as radio type, communications parameters, and so on.

If received signals are garbled, check things in this order, asking (and answering) each question before moving down the list:

1. **Is the signal tuned in properly?**

2. **Is the signal to the computer or data converter too loud or too soft? Try adjusting the signal levels.**

3. **Is the decoder software set to the right signal type?**

If you are transmitting data when the problem occurs, ask the receiving station on the other end these same questions. Check all of your transmission settings (the manual that came with your software covers how to do this).

What to do if you still have a problem

The troubleshooting information I've provided in these sections covers about three-quarters of all signal sending and receiving problems. The remaining quarter is split about evenly between two possible culprits:

✔ *Pilot error:* This is a polite way of suggesting that somewhere along the line you goofed. You can detect pilot error by having someone else check all the settings. Sometimes you need a fresh set of eyes. At some point, your friend will ask, "Why is this set to [fill in the blank]?" At this point, your ears will begin to turn red as you realize what the problem is.

✔ ***SDTs:*** SDT stands for *some darn thing*. An SDT can be a problem caused by some non-obvious, semirelated failure. SDTs can be harder to track down than pilot errors, but when you do figure out the issue, you will have a good story to tell!

Chapter 21

Troubleshooting Your Radio

*I*nto each life a little rain must fall, and in Radio Land, the rainmaker is a character by the name of Murphy. He spares no one, not novice, nor master. The successful radio maven is just better at giving Mr. Murphy fewer opportunities to visit and sending him on his way quickly when he *does* visit. In this chapter, I start by giving you some broad guidelines to help you identify the nature of typical radio problems. Then I run through some of the common causes and cures of those pesky Murphy bites.

Although I use the word *radio* in the following discussion, the instructions in this chapter follow the basic rules of troubleshooting. That means that regardless of *what's* malfunctioning (your computer-radio interface, as explained in Chapter 20, your radio, or your washing machine), you can use these troubleshooting techniques to identify and solve technical problems!

Hunting the Wily Mr. Murphy

Good troubleshooting is really very simple. You don't have to be an expert, but you do have to be patient, observant, and willing to take things one step at a time. There are very few problems that require serious sleuthing at the level of Sherlock Holmes. The key to good troubleshooting is following an orderly process and not jumping to conclusions. If you can do these two things, you'll be amazed at how well you can find and fix problems.

Keep a pad of paper handy when you troubleshoot. Start by writing down the symptoms. Then, as you take each step, make short notes to yourself. If you suspend the troubleshooting during a break, write down what you were doing when you stopped and what you think you should do when you resume. As any teacher will tell you, taking notes helps you maintain your train of thought and keeps you from repeating steps.

The following procedures are very, very general because there are so many types of radios and installations out there. You may have to adapt troubleshooting questions to match your particular circumstances and equipment. The manufacturer of your equipment probably has provided some troubleshooting advice in its operating and service manuals. Also, many manufacturers provide information on their Web sites, so be sure to read those, too.

Presumably, you're reading this chapter because your radio isn't acting the way you expect it should. Maybe the problem is innocuous, but maybe it's more serious, too. Voltages in ac power circuitry can be lethal and even modest amounts of RF power can give you a painful, if nonlethal, burn. Use your head! Only open up your equipment if you are familiar with safe technical practices.

Prosecuting Power Problems

Perform the following checks if your radio does not respond to the power switch or switches itself on or off at inappropriate times.

Don't fool yourself by saying, "Why, of course, the power's okay! I'm sure it's plugged in." During my experiences as a stereo repairman, I went on a surprisingly high number of service calls in which there was nothing really wrong with the stereo; a power cord that had become unplugged. The cat must have done it (of course!), but it's expensive, nonetheless, to have a service person make a house call to plug in the cord. Swallow your pride. Check power by doing the following:

1. **Confirm that the power cord (ac or dc) is connected firmly to the radio.**

2. **Confirm that the power cord is connected firmly to the power source.**

   The power source could be an ac socket in the wall, a dc power supply, or a cigarette lighter socket in the car.

3. **Confirm that the power source is live.**

   To test ac power, use a lamp to confirm the socket is supplying power. Similarly, to test dc power, use a light bulb or your voltmeter.

I don't recommend using another radio when testing a power supply because if the problem lies within the supply, the replacement radio might be damaged by a power surge!

If you're testing power in your car, use a 12 V light bulb or voltmeter. A weak car battery may show some voltage, but not enough to let your radio turn on. Excessive voltage can also cause trouble.

4. **Check the equipment fuses or circuit breaker.**

If the fuse has blown or a circuit breaker has been tripped, give your radio the *sniff test* before replacing or resetting it. If a failure in the equipment caused the fuse or circuit breaker to be overloaded, the characteristic acrid odor of overheated electronics is a common side effect. If you smell that smell, go no farther because the radio needs to be checked out by a technician. If there's no smell, replace the fuse or reset the circuit breaker and try again. If the radio springs to life, keep a wary eye on it for a while. If it fails again, note the circumstances.

5. **Check the power cord by either replacing it with an equivalent or checking it for continuity with a voltmeter.**

You can also use the voltmeter to measure voltage at the radio end of the cord while it is unplugged from the radio.

Be very careful when doing that test! Don't let the voltmeter probe tips touch each other while in contact with the cord (they'll short out the ac circuit with a bang) and don't you touch them — you'll be shocked!

6. **Make one last check of the power switch and any other controls that affect power, particularly back panel switches or cables that affect how power is turned on and off.**

Your owner's manual and the manufacturer's Web site can be very helpful when checking out the back panel and how the radio is turned off and on.

7. **If you've done all these things and the radio is still not powering up, skip to the section "Visiting the Radio Doctor."**

Do not pass Go; do not collect $200. Your radio is probably sick and needs a technician's attention.

Anticipating and preventing ac power problems

Don't mess with alternating current (ac) power. A flaky power cord or wall socket is a surefire road to, guess what? A fire! Replace cords whenever you can. Power cords are so cheap that there's no reason to repair a cord unless it's designed to be repaired. Old cords, such as the one attached to Grandma's bedside radio, often have brittle and cracked insulation, and should be replaced regardless of whether they are currently causing problems. Power strips with multiple outlets are far superior replacements for three-way taps or multiplug extension cords. They're also quite inexpensive.

Over time, older surge protectors can gradually became more and more like heating elements. They eventually get hot enough to damage the power strip or even start a fire. You can feel the heat from a failing surge protector on the outside of the power strip, usually near the power switch or where the cord is attached. Plastic power strips may be discolored or warped near the hot component. Throw away old power strips and get new ones.

Dealing with dc power problems

Batteries are by far the most common source of direct current (dc) voltages for radios. That's why I dedicate a whole chapter to them (Chapter 19) in this book! Chances are that if your battery-operated radio isn't working and you suspect a power problem you need to replace the batteries.

Of course, allow me to recommend purchasing a battery tester, such as the one kept conveniently in a hall closet. A battery tester is a good thing to have around the house; because everyone tends to have so many battery-operated things, you'll use it for a lot more than the batteries you use for your radio.

A weak battery might look fine to a voltmeter because very little current is drawn. The battery tester loads the battery to be sure of its current supplying abilities.

If you use a nonbattery dc power supply, the output voltage should change very little whether it has a radio attached or not. Some power supplies can output too high a voltage when the *voltage control circuit* fails. This is very hard on attached equipment. If too heavy a load is attached, say a 25-watt radio to a 1-amp supply, the output voltage will drop or *sag* to a low level. A typical symptom is dimming panel lights or indicators when transmitting. Sagging voltages can damage both the radio and the supply, so be sure the supply is rated to supply enough current. A 60-cycle or 120-cycle *hum* or tone on your signal is also a symptom of an overloaded or failing supply. If any of these fit your symptoms, replace the supply at once.

In a car or boat, even with a fresh battery, the voltage available varies quite a bit. Without the engine running, battery voltage is somewhere between 11.5 and 12.5 V, depending on battery condition and charge. Starting the engine causes the voltage to drop as low as 8 or 9 V because of the exceptionally heavy load the starter motor presents. A running engine's alternator can drive the voltage to between 13 and 15 V. As with the ac line, glitches and spikes are common and can be substantial. Equipment not designed for automotive use often suffers from mysterious failures because of these wide excursions in voltage.

Higher-power radios, such as marine VHF transceivers, draw plenty of current (as I discuss in Chapter 18). In the outdoor environment and subjected to the vibration of a vehicle, connections can easily become loose or corroded. A small amount of extra resistance, even a fraction of 1 ohm, can cause a significant voltage loss in a high-current connection, getting hot in the bargain. If you suspect low voltage, double-check all your power connections. A discolored terminal is a sure sign of overheating and should be replaced. An underrated switch can cause the very same problem.

Whenever you are working on a battery-powered radio, be sure to have a fuse in the power lead of the radio. Fusing the negative lead is a good idea, too (see Chapter 18). Batteries store a lot of energy that can burn up a radio or start an expensive vehicle fire. A short circuit can melt connectors and wires, burning whoever touches them, as well. Don't fall into the "it's only 12 volts" trap!

Solving Operating Problems

If you have ascertained that power is flowing to and from the radio, but the radio still doesn't seem to be operating properly, start by checking your operating manual and the manufacturer's Web site for troubleshooting guides. If those aren't available or aren't helping, the following steps can help you isolate what's wrong.

1. Set every control, front panel and back, top, bottom and sides, to a known and acceptable position.

Often, the owner's manuals for more complex radios contain checklists for the main controls just for situations like this.

Start by following those steps. If you don't have a checklist, go over every control or switch and be sure it's set to a setting that won't cause the symptoms you are observing. You can easily (and unknowingly) bump or rub small knobs and buttons into unintended positions.

2. Check the squelch settings.

A very common problem is that a radio seems dead on receive because the squelch control has been turned up too far. Often, the squelch control is paired up with a volume or AF gain knob, so accidentally changing the squelch setting when adjusting the volume is a fairly common problem. Set the squelch to the off or full-open position before declaring that the radio is broken.

Also check to see that the transmitter is not locked on because the microphone PTT switch is stuck or a switch is set to Transmit. (Some radios have a button labeled XMIT (transmit) or MOX (manually operated transmit). If this is on, the radio transmits forever!)

Check, testing one, two. Check.

Some days, I think the most common phrase on the airwaves is, "Can I have a radio check?" People just like to know their equipment is working. Your service may have specific rules about conducting over-the-air radio testing. The most common method is to press the mike switch and say, "[Your identification] requesting a radio check, please." Then wait. If you get no answer in five to ten seconds, try again. If you still receive no response, quit or try a different channel.

You may get a simple reply from another station, such as "[Your identification], this is [the other station's identification] and you are loud and clear." Other times, you may hear a reply like, "[Your identification], this is [the other station's identification], and you are weak and scratchy."

(Be honest when responding to a radio check.) Assuming you don't need any further information, reply, "Thank you, this is [your identification], clear (or out)."

Unidentified transmissions are frowned on. Say you transmit without speaking into a repeater to see whether you can hear the repeater's transmitter when you let up on the mike switch. This form of unidentified transmission is called *kerchunking* because of the annoying sound other suffering repeater users hear as the repeater turns on and off. If you have a reason to believe your radio (or the repeater) isn't working, you should simply say, "[Your identification], monitoring."

3. **Temporarily remove all accessories and connecting cables, except the power supply and the antenna, to see whether the radio operates normally with them removed.**

   If so, reinstall the accessories one at a time to see when the problem reappears. This strategy helps you isolate the culprit. Pay attention to signs of intermittent operation of the radio when you remove and insert cables.

   Problems in tuning the radio can often be traced to a Lock control that has been inadvertently activated. These controls are often in obscure locations and are easy to overlook. Oopsie.

4. **If the radio is not receiving signals, remove the antenna or antenna cable and listen for crackles and pops as you disconnect it.**

   Most radios give an indication that something is happening out there as the antenna is disconnected and reconnected.

5. **If the radio doesn't seem to be transmitting, check the microphone PTT switch. Then check the output power indicator or meter to see if the transmitter thinks it's putting out power.**

   Almost every radio gives some sign (usually a light or other indicator) that the transmitter is on. Battery-powered radios may not transmit if the batteries are too weak, giving some kind of low-battery indication.

If you have a power meter, now is the time to use it. Finally, perform a radio check with another nearby user.

6. **When all else fails, there is often a way to reset the radio's control circuits.**

Find out more information in the radio's operating manual. You may discover that you have several reset options, including a *soft reset,* which is similar to rebooting your computer.

A full, or *hard reset,* may erase all the radio's memories, so don't do this until you've exhausted other choices.

Banishing Noise

Noise has been with humankind since the dawn of time. It's been with radio users almost as long. Even Marconi writes of having to overcome noise. Everyone with a radio eventually encounters some unwanted noise, so finding out how to deal with it effectively is very useful.

When radio operators refer to *noise,* they're talking about what most people call *static.* Whatever you call it, it's characterized by random popping, crashing, buzzing noises that interfere with reception that *aren't* signals from another radio.

Noise has many sources, some of which you can avoid or even do away with. Although your radio may have noise reduction features, such as a *noise blanker* or *noise limiter,* the best thing by far is to eliminate noise at its source. Here's the threefold response to nipping noise in the bud:

1. **Identify the type of noise.**

2. **Track down the source of the noise.**

3. **Eliminate the noise at the source.**

Neutralizing power line noise

A common source of noise is from the ac power lines. Power line noise has a characteristic buzzing or humming with a frequency of 60 or 120 Hz. It sounds like a very large, angry bee or wasp and you can hear it whenever you drive under a high-voltage power line while listening to an AM broadcast station. Power line noise is usually a problem when trying to receive an AM signal, such as shortwave broadcasts, marine SSB, aviation communications, and TV signals.

Power line noise can be steady or intermittent and it may change with the weather. It's usually caused by dirty or defective hardware on a power pole. You can find the source of power line noise with an AM broadcast receiver (carry it around with you or use the one in your car to pinpoint the problem line). Tune to a frequency between stations where no broadcasts are present. If you're close to the noise source, you'll hear the buzzing sound. If not, you'll hear the faint crackle and hiss of normal atmospheric noise. Walk or drive around your neighborhood while listening to the radio. When you get close to a noise source, the buzzing will get louder. A strong noise source may be audible for a block or more, so you'll have to move around in order to find the very strongest location.

Assuming the strongest location is near a power pole, ***do not*** kick, shake, or climb the pole or guy wires! Write down all the identifying numbers on the pole. Every utility that uses power lines and poles — electric, telephone, and cable TV — identify every pole they use. The electric company uses one of those numbers (or sets of numbers). Call the utility and report that you are experiencing interference that you believe is coming from a specific pole. Provide the address or physical location and the identifying numbers. The utility will *eventually* send a line crew to inspect the pole. However, there's no guarantee that the crew will find the noise immediately or that it will be completely eliminated.

Noise coming from substations and high-voltage lines is much more difficult to deal with. If you live close to such installations, you are probably out of luck in getting rid of the noise completely. In this situation, you have to rely on the radio to work around the noise as much as possible.

Angling around appliance noise

Appliances and gadgets that operate on ac power generate an unbelievably common source of noise. Any motor with brushes, such as those used in blowers and appliances, makes sparks that generate radio noise. The frequency of the noise depends on the speed of the motor.

Switches (such as thermostats) that turn heating elements on and off generate noise from a small *arc* on opening and closing, particularly when their contacts are worn. The arc also heats the contacts and can become a fire hazard, so replacing switches provides more benefits than just a quiet radio! Light dimmers that work by cutting out a little bit of each ac cycle to reduce the power to lamps also generate noise.

Appliance noise sounds like power line noise, but usually switches on and off in some kind of pattern that is a clue to the type of appliance. For example, a noise that runs steadily for a few minutes, and then turns off for a few minutes, and then returns again usually comes from a furnace or air conditioner motor. Noise that runs for short periods in no particular pattern is probably caused by a household appliance, such as a sewing machine, blender, electric drill, hair dryer, or similar appliance. A regular pulsing might be caused by a neon sign or blinking light. Identifying these noise sources, which can be heard up to a couple of blocks away, can require some detective work.

To find a noisy appliance in your own home that doesn't correspond to an audible sound such as a motor, wait for the noise to appear and then turn off the circuit breakers one at a time until the noise suddenly stops. If the noise doesn't stop; then the noise is either coming from the same circuit as your radio or is not coming from your home.

To get rid of these noises, you can often apply an *ac line filter*, such as the examples shown in Figure 21-1. You can install an external filter between the appliance power cord and the ac wall socket. Some filters attach to the appliance. An internal filter requires rewiring the appliance's power connections, so I don't recommend that you try it yourself unless you are competent to do so.

Figure 21-1:
These are examples of ac line filters that keep noise from being transmitted by power cords and house wiring.

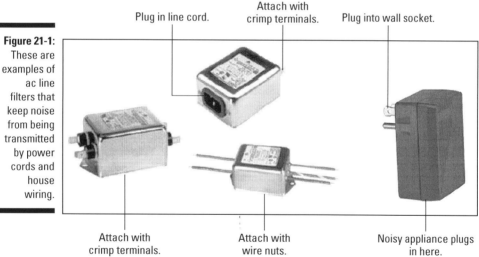

Plug in line cord.

Attach with crimp terminals.

Plug into wall socket.

Attach with crimp terminals.

Attach with wire nuts.

Noisy appliance plugs in here.

Nullifying noise from engines

The primary source of noise in a RV, car, or boat powered by a gasoline engine is *ignition* noise. The spark plugs, even though imbedded in the engine's metal block, still generate a powerful pulse of energy each time a cylinder fires. Distributors and ignition points also contribute their own sparks. It's not hard to identify ignition noise. At idling speeds, it sounds like a steady "pop, pop, pop" that becomes faster and faster as you accelerate.

The engine's high-voltage wiring acts as an antenna for the noise, which can also find its way onto the vehicle's dc power distribution wiring. It's no wonder that these pulses find their way into your radio! Begin your fight against ignition noise by reducing how much of it is generated. Use spark plugs with built-in resistors (called *resistor plugs*) that limit the amount of energy in the spark. Be sure all of your high-voltage cables are in good shape, and also check the distributor, points, and condenser. Electronic ignition systems should be checked for good connections, particularly the grounding.

Other common noise sources in vehicles are the battery charging system and electrical motors. Batteries are usually charged by an alternator or generator driven by the engine. Noise from these sources is a reedy-sounding tone known as *alternator whine* that varies with engine speed. There are many electric motors in a vehicle, as well. Fuel pumps, electric windows, and windshield wipers are just a few of the common vehicle noisemakers. The first line of defense against these is a noise filter. If the noise persists, consult the vehicle dealer for possible solutions. Some manufacturers have service bulletins regarding radio noise problems.

Attacking atmospheric noise

Atmospheric noise plagued Marconi and bedevils radio enthusiasts to this day. Even in the complete absence of human-made noise, natural processes generate noise. The steady crackle of noise between broadcast stations and competing with weak signals comes from many sources, such as storms and the sun illuminating the ionosphere. There isn't much you can do to reduce the noise itself, of course. Your best bet is to purchase a radio with noise reduction technology or put up antennas that reject signals and noise from directions away from the signal you're trying to listen to.

Dealing with Interference

Different than noise, *interference* is generally understood to refer to unwanted signals or the effects of other signals. For example, ham radio operators refer to interference as *QRM* and atmospheric noise is *QRN*. Interference is categorized in two types: the kind you experience from others and the kind caused by you when you transmit.

Received interference

The first thing to do when you are receiving interference from other radio signals is to be sure the problem is not generated by your own radio. Modern radios have very good receivers, but can be operated in ways that make them susceptible to interference or even generate false or *spurious* signals that seem like interference from another station. Before getting hot under the collar and marching off to find the offender, check these possibilities first:

- **Overload:** Receivers can only withstand so much received signal strength before being overloaded. If you live near a strong transmitter, you may experience overload. This is particularly the case with consumer models of stereos and televisions. When this overload occurs, the receiver's output becomes distorted and erratic. The strong signal doesn't even need to be in the same frequency band to which the receiver is tuned!

  The resulting interference is not the fault of the transmitting station at all. The best solution is to filter out the strong signal. There are too many different circumstances to give specific solutions, but Table 21-1 provides some guidance.

  If your radio has an *attenuator*, which reduces incoming signal strength, try using it to see if you can reduce the strength of the overloading signal. Although it also reduces the signal you want to hear, that often isn't a problem, even at lower signal strengths.

- **Preamplifiers and noise blankers:** A preamplifier or *preamp* dramatically increases a receiver's sensitivity to weak signals. However, if a strong signal is present, the preamp creates overload problems. Turn the preamp off and if the interference goes away, you'll know that it's an overload problem. Noise blankers often cause spurious interference, too. They work by detecting the sharp, short pulses characteristic of most noise, and then muting the receiver for that period of time. The trouble is that they often can't tell the difference between a strong signal and noise. That means they mute the receiver inappropriately; the result sounds like interference. Turn off the preamp and noise blanker before you determine whether you're receiving real interference.

- **Intermodulation:** Intermodulation, or *intermod,* is a 50-cent word that identifies what happens when two signals get combined in a way that creates more signals. The additional signals are called *products*. One or more of the products might appear right on the channel you're trying to use. They sound like actual signals to your receiver because they are signals. They come and go along with the signals that generate them. If the combining happens in your radio because of overload, the solution is a band-pass filter or some attenuation of the incoming signals. If the signals are being generated elsewhere, you can't do much except move to a different channel.

- **Adjacent channel interference:** A strong signal on a nearby channel can cause interference if your radio is not *selective* enough to reject it in favor of the one on the channel you're tuned to. In some cases, your radio's manufacturer may have an extra-strong filter that you can add to your radio that improves the rejection of nearby signals. If you often find yourself listening to a crowded band, these extra filters are often a good investment.

Table 21-1	Interference-Reducing Filters	
Filter to Use	*Effect of Filter*	*Type of Interference*
High-pass	Rejects signals below the *cutoff* frequency.	Use on FM and TV receivers experiencing interference from shortwave, AM, amateur HF, and CB transmitters.
Low-pass	Rejects signals above the *cutoff* frequency.	Shortwave and amateurs use for interference from strong local FM or TV transmitters. Use on FM and TV receivers receiving interference from commercial VHF and UHF transmitters.
Notch or Trap	Reject a specific range of frequencies.	Often used to reject signals from a single nearby paging, FM, or TV transmitter.
Band-pass	Rejects signals outside a specific range of frequencies.	Used when multiple interference sources are present and to reduce intermodulation from strong nearby signals.

Avoiding being the cause of interference

If you're not transmitting radio signals, you can leave the room. The rest of you, please stay behind. Just about everyone that uses a radio to transmit will encounter a situation in which transmitted signals cause interference. Interference you cause is more accurately referred to as *EMI* (electromagnetic frequency interference) or *RFI* (radio frequency interference). The symptoms may vary; others may hear a click or thump in an appliance; your voice may be heard in a speaker; or there may be a serious disruption in function. Because everybody will be looking at you as the culprit, here are some guidelines.

If you're operating in a licensed service and your transmitting equipment is operating properly and within the required power limits, you are usually not responsible for eliminating the interference. The FCC's rules are quite clear about this, particularly if the device being interfered with is an unlicensed receiver. These are called *Part 15* devices and they're required to accept any interference they encounter or to stop operating if they cause interference to a licensed station. Part 15 devices, including cordless phones, wireless network gadgetry, and so on, have this spelled out in the operating manual or on a sticker right on the equipment.

The top five most common radio problems and how to fix them

To get the best performance out of your radio when you transmit and receive signals, study the operating manual and follow instructions. If you're not getting as much out of your radio as you think you should, here are five of the most common problems and what to do about them:

✔ **Defective microphone:** Mics undergo a lot of mechanical abuse, which puts stress on connections. If the jacket of your mic cable looks frayed where it enters a connector or if it's loose in the connector shell, replace the connector properly. Another common mic problem occurs when mic elements that change the voice into electrical signals become clogged or cracked with time. If your voice sounds muffled or distorted over the air, check the microphone and cable.

✔ **Old or poor-quality antenna cable:** Cheap coaxial cable is no bargain. Pass up those great deals on slightly used cable. If the jacket is cracked or abraded and water gets in, signal losses go up dramatically. Good quality cable is available at reasonable prices from many vendors; don't scrimp on this important point in your system.

✔ **Poor antenna connections:** If you are using an external antenna, the electrical connections are likely exposed to the weather and vibration. When you install an antenna, be sure to protect the connections with a protective sleeve or *boot*, a good quality of

electrical tape (Scotch 33+ is excellent), or a conformable coating such as RadioShack 278-1645. At least once a year, check the connections to be sure they're not broken or corroded.

✔ **Poor radio signal connections:** Poorly wired connectors cause a great deal of grief. If you install your own connectors, find out how to do it right, using the proper tools and materials. Be sure that connectors aren't loose.

✔ **Power and grounding problems:** Be sure that cables are big enough to carry the expected load and that all connections are tight and clean. A good ground connection can also help avoid distortion of your voice from the transmitted signal being picked up by the microphone wiring.

If you're having problems with your radio reception, your first urge may be to turn all the gain controls to maximum and shout. This isn't going to resolve the problem — the resulting distorted signal will still be, well, distorted. Your radio's operating manual will show you how to adjust the transmitter for the clearest signal. Have a more experienced radio friend listen to your signal and help you find the right settings, the right speaking level, and right way to speak into the microphone. There's no reason to sound like a bad public address system over the air.

If the shoe is on the other foot and your unlicensed radio, such as an FRS handheld transceiver, interferes with a radio being used with some other licensed service, you are required to stop operating if asked. These rules were made to allow useful unlicensed devices to coexist with the more capable licensed radio users.

Causing interference is not a right or wrong issue. Just because you don't *have* to reduce interference under some circumstances doesn't mean that you shouldn't reduce interference when possible. The easiest solution you can apply is distance. Moving the transmitting antenna away (including higher or lower) from the affected item (or vice versa), makes the signal from the transmitter weaker and less likely to cause interference. A second solution is for the party receiving interference to install the appropriate filter in the incoming signal line (refer to Table 21-1).

Interference is not always caused by problems with a receiver. In many cases, the interference is caused by signals being picked up on speaker or power cables. This is called *conducted interference*. Conducted interference is difficult to resolve, but you may be able to make use of one of the following suggestions (examples are shown in Figure 21-2):

✓ **Direct current chokes:** You can install a dc choke inside equipment to prevent HF signals from getting in through the power connections. Installing chokes is a task best done by a technician.

✓ **Ferrite cores:** If an ac power source doesn't have a standard line cord, such as a wall transformer, winding several turns on a ferrite core, such as the RadioShack 273-104 or 273-105 may help. This technique can also be used on audio, video, and speaker cables.

✓ **Power filters:** In the section "Angling around appliance noise," earlier in this chapter, I tell you about ac line filters. These power filters also filter out unwanted RF signals and may prevent others from receiving interference when you're transmitting.

✓ **Telephone filters:** A telephone filter is inserted at the phone by disconnecting the phone line and reinserting it into the filter.

After carefully isolating the way in which the interference is being received by the affected device, you can choose the appropriate filter. The tried-and-true method is to disconnect everything but the power source and reattach cables one at a time until the interference is again observed. At that point, you have a much better idea as to how to get rid of the interference.

The *Interference Handbook* published by the American Radio Relay League (www.arrl.org) is an awesome and inexpensive reference, covering all kinds of interference problems and techniques for curing them.

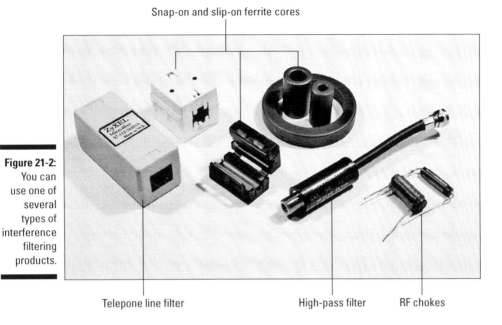

Snap-on and slip-on ferrite cores

Figure 21-2:
You can
use one of
several
types of
interference
filtering
products.

Telepone line filter   High-pass filter   RF chokes

Visiting the Radio Doctor

Oh dear, you've tried everything and still no luck; or worse, there's a faint odor of fried electronics lingering in the air near your radio. It's time for a trip to the radio doctor. If your radio is still under warranty, return it to the manufacturer. Otherwise, look in the Yellow Pages or on the Web for a local radio repair shop that is considerably less expensive than the manufacturer's rate. Call the shop to be sure it can repair your radio. Also see whether the shop has references you can check.

You can save yourself some time and money if you slow down and follow these steps:

1. **Write down everything.**

   Include the radio's symptoms, along with your name, address, phone number, and e-mail address. Include the radio's model number and serial number.

   For example, you may write "Radio is locked in transmit with or without the microphone attached, smells hot." Or perhaps, "Others hear my voice as distorted and broken even though the signal seems strong." Writing down the symptoms helps the technician zero in on the problem quickly, saving labor costs.

Don't guess or jump to conclusions — you don't want to lead the repair tech astray! Avoid writing notes such as, "Radio locked because the power supply must be defective." You may cost yourself an hour of labor before the tech finds the *real* problem.

2. **Contact the dealer or manufacturer to start the return process.**

   You may have to get an RMA (Return Materials Authorization) number or a repair tracking number before sending equipment back to the manufacturer. You may have to send the radio to a special address.

3. **Make a copy of your symptom list with your name and address, clearly show the maximum expense you are authorizing without being contacted for approval, add the RMA or tracking number to the list, and attach the list to the radio with tape.**

   A local shop may also provide an estimate before proceeding with the repair.

4. **Protect the radio.**

   Start by wrapping it in a plastic bag or even in plastic wrap. This extra plastic covering keeps packing material out of the little nooks and crannies of your radio. Include any accessories required to operate the radio. If you're not sure what to include, call the manufacturer or repair shop to ask whether to include the power cable or cord and the microphone.

   Even if you're hand-carrying the radio to the shop, be sure to pack it in at least one box for protection and tape your list of symptoms to the radio.

5. **If you're mailing your radio, pack it in a shipping carton and send it away.**

   Use the original shipping carton and packing materials if you have them. If not, get a box big enough to contain the radio plus at least a couple of layers of shock-absorbing material, such as Styrofoam packing pellets. Then *double box* the radio by placing the first carton inside a second box with a couple of layers of packing material between the two boxes.

   Tape the outer box carefully and show the RMA or tracking number prominently near the address. Ground shipment is fine and often available through shipping stores in malls and shopping centers. Insure the radio for its replacement cost.

By following these steps, you can reduce wasted effort by the repair technician and minimize the chance that your radio will be lost or separated from your name and the symptom list. These will help you get a repaired radio back home and on the air as quickly as possible.

Part V
The Part of Tens

The 5th Wave   By Rich Tennant

"The only thing I'm worried about is where to mount the antenna."

In This Part . . .

Some say The Part of Tens is the best part of a *For Dummies* book. Here you can find a list of secret knowledge, guaranteed to launch you into radio orbit. Then you can find ten tips for those adrenaline-filled moments when you use your radio in an emergency. What if the radio itself needs help? Look no further than Part V for some quick-and-dirty radio first-aid pointers. Seriousness aside, the part concludes with ten sure-fire ideas to make using your radio fun.

Half the fun for radio lovers is passing along radio secrets to readers just like you. Take advantage of all the resources you can, and maybe we'll meet up on the airwaves someday! As the hams say, "73," or, "Best regards!" See you down the log!

Chapter 22

Ten Radio Secrets

*N*o matter what kind of radio you're using, common tips apply to them all. Finding and following these secrets will imbue you with radio smarts that keep you on the air! So enjoy the spirit of the tens in this short-but-sweet summary of the most important radio secrets you ought to know.

Listening Rules

By using your ears instead of your mouth, you can discover a tremendous amount of information about operating and using your equipment.

Keep quiet to find out what sounds good and what doesn't, how radio signals get around, and all sorts of amazing things. It's strange but true that listening makes you a better talker. Your kindergarten teacher was right.

Talking Louder Doesn't Do You Any Good

Lung power won't push your signal any farther or get your message through any better. Modulate your voice, enunciate clearly, and sound like a pro every time. Don't you wish everyone would follow this rule in real life? If other stations continue to have trouble understanding you, it's a good idea to check your microphone — see Chapter 21.

It's All in the Antenna

A dollar spent on a good antenna goes farther, so to speak, than on anything else you spend money on in your radio hobby. No other component of radio makes a bigger contribution to successful contacts.

Use high-quality antennas and feed lines and keep them in good shape. Your signal will thank you for it.

The FCC Does, Too, Care

Some people may try to get you to believe otherwise, but the FCC does care what you do. Don't believe it when your hear, "You don't really need a license. You won't get caught because no one ever checks." Those are famous last words! *Bootlegging* (that is, transmitting and receiving signals without a license for the service) causes interference and *does* attract the attention of the FCC.

Even if the FCC *didn't* care (did I mention that it does?), unlicensed users make things more difficult for everyone. Personal licenses are not expensive and help keep the airwaves useful for everyone.

Setting Up Your Radio Correctly the First Time

Resist the urge to hurry up your installation. Don't skip a recommended step or hold off on installing a part just because you want to be on the air right away.

Radio signals are easily leaked away or contaminated with noise and interference. An erratic power supply causes all sorts of trouble. You'll save a lot of time in the long run if you plan and set up your installation smartly.

Don't forget to talk to a radio or electronics guru if you don't know a lot about electronics. Talking to an expert can not only save you time, but it can also ensure your safety.

Finding the Hot Spot

In the VHF and UHF bands, reflections and multiple paths can create *hot spots* and *dead spots* that are only a few inches apart. If you find making contact difficult, try moving your radio 6 inches to a couple of feet from where you originally were. Solid contacts may be within a step or two!

Planning for the Worst

Think ahead to foil Murphy's plans! The following list offers a few of the simplest things you can do to avoid being without your radio when you most need it:

- Have spare batteries on hand.
- Keep checklists (and follow them).
- If you're using portable radios, make sure that everybody on your team has the same plan and a backup plan.
- Regularly maintain your radio. Conduct radio checkups often.

All these tips can keep Murphy from sneaking up on you.

Getting What You Pay For

Be wary of shiny gadgets and amazing performance claims from manufacturers you've never heard of. Look for a manufacturer with a complete line of radio products and a solid warranty.

A portable radio should feel like it can withstand being dropped, because you will drop it at least once! A cheap radio is no bargain.

If you have any question about whether a product, attachment, or piece of software is of high quality, check with others in the radio community, read online reviews, and ask salespeople at your local electronics store. Avoid making decisions in a vacuum and recognize that you may have to pay a little more to have the quality you need.

Following the Ten Count

Sooner or later you'll hear or be tempted to engage in a radio squabble over who was using the channel first or who's interfering with whom. Don't go ballistic; just figure out how to work around the problem. A shouting match over the airwaves won't solve anything.

Increase the peace. As you discovered as a kid, counting to ten can keep you from saying something you regret. This simple action can help you avoid a lot of unhappiness. Life's too short.

Discovering the Best Way to Relax

There's no need to tense up over speaking over a radio. Take a deep breath now and then to keep your voice sounding natural and clear. Don't worry about making a mistake — we all do at first. Laugh it off and grow. Pretty soon you'll be helping another newcomer to the airwaves.

Keep your transmissions short. Leave yourself time to think. Remember to enjoy the magic of radio!

Chapter 23

Ten Emergency Tips

*N*o one hopes for a catastrophe, but stuff happens. Maybe there's a storm, earthquake, or other natural disaster. Or perhaps someone in your party is lost or hurt. Chapter 6 talks about communicating in those circumstances. Regardless of what kind of radio you have or the reason you bought it, knowing the simple, but powerful, tips in this chapter can help you communicate in an emergency or disaster situation. Use this chapter for the short version of the top ten things you need to know.

Using Your Radio Regularly

The old Scouting mantra to be prepared works even better on the radio. Funny thing is that for all my talk in this book about batteries and antennas, the most essential tool you need to stay prepared is you. If you know what you're doing, you'll be calm, cool, and collected in an emergency situation.

The best way to be prepared is to use your radio regularly. Regular use will give you that nice feeling of knowing that you can make contact whenever you want to. The time you spend in preparation will repay you many times over when you need it most.

Getting Yourself under Control

When the adrenaline is surging, it's easy to panic. Panic is not good. You have to be coherent and clearheaded to communicate by radio.

You may feel a need to communicate *immediately* if things are dire, but don't press that microphone button until you can think clearly. Otherwise, you'll waste precious time and possibly give out erroneous information.

Supplying Just the Facts

Stick to what you know! Limit the information you supply to the good, old *who, what, where, when, why,* and *how* questions. Do not guess or assume anything. Speculation on your part may be overheard and taken as fact by others. Rumors spread like wildfire — don't start them!

Embellishing information is easy to do when you're trying to help. Instead of speculating, you should hone your observation skills. When camping, hiking, or doing other outdoor activities, pay particular attention to your surroundings, carry a map and compass, and make note of landmarks.

Being Smart with Resources

If you're running your radio on emergency power, keep transmissions short. Save your batteries by turning the radio completely off when you can. Stick to any prearranged schedules and channels to communicate. Keep volume low and don't leave the squelch or monitor function open.

Knowing Where to Tune

Find the emergency channels or frequencies that are used in your area. Memorize them, or better yet, write them on something to keep in your pocket, car, or emergency kit. Get in the habit of listening to emergency channels from time to time in case someone else is in need of assistance.

Taking Advantage of Geography

Extend your range and maximize your signal strength by taking advantage of height. Drive to a hilltop or walk to an upper floor or an upstairs bedroom. Go to a clear area, if you can. Remember to move out of dead spots and away from foliage that might soak up your signal. In an emergency your signal needs to be as strong and clear as possible.

Writing Everything Down and Keeping It Short

When you are relaying messages, such as damage reports or personal health and welfare, get information as close to 100 percent correct as possible by writing all details down. That way you won't have to guess or remember. If possible, use a form to help guide your thoughts under stress. Remember to note who the message is from and get complete delivery information. (The ARRL's Radiogram form at www.arrl.org/FandES/field/forms can serve as a guide.)

Following the Plan

Make sure that before an emergency arises you have contingency plans in place and that everyone knows what the contingency plans are. Everyone should know that no matter what happens no one should deviate from the plan. Following your group or community communications plan is essential to getting and staying organized! Be on time for scheduled communications and make sure you're on the right channel. Operating rules and plans become extra important under emergency conditions because they can save time and lives.

Using the Buddy System

Having a radio buddy with you works wonders, both in avoiding emergency situations and working your way out of them. A radio with a ranging system can help you stay linked. Even at home, practicing and training goes so much easier with a friend.

Practice, Practice, Practice

There are many ways to hone your skills. Rehearsing your plans with family and friends is a great way to start. Take online courses offered by manufacturers, organizations, and public safety agencies such as those at www.fema.gov (click the Education & Training link). These courses are often free! Above all, practice using your radio.

Chapter 24

Ten Radio First-Aid Techniques

As with any gadget, there are lots of reasons a radio could cause you to say, "Uh, oh!" Maybe you dropped it! Maybe the dog chewed it! (Who knows?) At any rate, if the radio's not working, here are some field-fix techniques that can get you back on the air . . . for a while, anyway. The ten temporary solutions offered in this chapter can keep you going until you can get the radio to the manufacturer or a licensed technician.

Resetting the Radio

A radio that's been dropped or abused may be confused. Most radios have a reset function that *reboots* the radio — similar to the Ctrl+Alt+Del function used by PCs.

If your radio is acting odd and other attempts at repair fail, perform a reset to return the radio to its factory configuration. Your owner's manual contains specific instructions on how to reset the radio. The usual methods involve either holding down combinations of keys when turning the radio on or pressing a hidden switch. Resetting the radio will probably erase all memories, so reset wisely!

Replacing Lost Antennas

If your antenna snaps like a twig your reception *will* be affected, and not in a good way. You can usually bet that sticking a piece of wire into the antenna connector will help you out to some degree. Try 6-inch wire on an FRS/GMRS radio. VHF radios, such as MURS or BRS, work best with a 17- or 18-inch wire. CB radios need a much longer piece of wire — one that's 8 feet, 8 inches long.

Putting the Wrong Batteries to the Right Use

If you have the wrong size batteries, you can use tape to connect the batteries end to end (+ to -) and then use scrap wire to connect the ends of the stack to the contacts in the radio. Be sure to follow the + and - markings in the radio. This ain't pretty, but it works.

Fixing a Faulty Pushbutton

When a pushbutton stops working, try pressing it with the point of a pencil or ballpoint pen at different angles and locations. The contacts may be worn just in the spots where a finger causes them to come together.

Working around a Broken Speaker

If your radio accepts headphones, you can replace a broken speaker with small speakers — the sort used by portable CD and MP3 players. This isn't an ideal solution, but at reduced volume, these speakers work quite nicely.

You can also run sound to your laptop computer if you have one. The laptop's sound card microphone input accepts audio output from radios, too.

Splicing Together Torn Wires

Cables for speaker mics, headphones, and power supplies often break where they enter a housing or connector. Use a sharp knife to strip the insulation away and then twist the wires back together for a temporary fix. Headphone and microphone wires may actually be tiny coaxial cables with an inner and outer conductor, so be sure to separate them.

Working through Wind and Noise

If you're using your radio in the elements, wind and other noise can mask your voice on a built-in microphone, making you impossible to understand. Try operating with a jacket or bag over your head. Don't yell; yelling simply makes the problem worse. Have others surround you to block wind and absorb sound.

Rescuing an Immersed Radio

If your radio falls into a river, lake, ocean or bathtub (I won't ask), you may not need to kiss it goodbye. Minimize damage by following these steps:

1. **Immediately remove the batteries, and if you can, the back of the case.**

2. **Flush the radio with fresh, clean water.**

3. **Shake the radio dry and drain it.**

4. **Operate all the knobs and buttons to clear debris from the contacts.**

5. **Place the radio in a warm spot for several hours to dry.**

   Good luck!

If your radio has gotten wet but not completely dunked, you only need to follow Steps 3 through 5.

Building an Emergency Charger

If your rechargeable batteries are kaput, you can partially recharge them by using a car battery and a dome light bulb to limit current. Of course, if you have the car at your disposal, you may want to simply drive it to the hardware store and purchase a new charger. On the other hand, if you're in the wilderness somewhere, these steps come in handy:

1. **Connect individual cells in a string.**

   See "Putting the Wrong Batteries to the Right Use," earlier in this chapter.

2. **Connect the car battery's + terminal through the light bulb to the + battery terminal and both - terminals directly together.**

3. **Make the connection for a few seconds at a time.**

   Don't let the batteries get hot.

4. **Try the batteries after a dozen or so connections.**

Making Do with the Tools around You

What if you don't have your tools handy? A multipurpose pocketknife often has plenty of options and is a good tool to have on hand in its own right. Otherwise, you can press a nail clipper into service as a wire cutter and stripper. The metal nail file in the clippers makes a nice screwdriver. You can also use clingy plastic wrap or postage stamps rather than tape.

Chapter 25

Ten New Ways to Have Fun with Your Radio

*P*eople use radios to support a lot of activities, making them safer, more efficient, and just more fun. As you become more adept at using your radio, you'll think of many new ways to put it to use. This chapter, in true Part of Tens form, offers some (ten) such ideas, a few of which are pure radio.

Direction Finding

Radio direction finding is a sport activity increasing in popularity around the world. This activity combines *orienteering* (the activity of finding your way over land by using a map), radio, and personal fitness. There are even world championships! Hidden transmitter hunts (also known as *fox hunts* or *bunny hunts*) are similar activities. All you need in order to be a part of the action is a handheld receiver (scanners often work fine) and a lightweight directional antenna. Oh, and your wits!

Joining in a direction-finding competition can be a great way to make friends. It's also a great way to increase your radio knowledge. Here are a couple of Web sites that tell you more about this exciting activity. They list the specific equipment you'll need:

- **Galaxy Directory listing for Radio Direction Finding:** www.galaxy.com/ galaxy/Leisure-and-Recreation/Radio/Amateur-Radio/Radio-Direction-Finding
- **Amateur Radio Direction Finding (ARDF) Web site:** www.ardf.us

Hilltopping

For an interesting day trip, take a portable or mobile radio and maybe a lightweight directional antenna to a high spot with a clear view of a populated area and try it out. Hilltopping is a great way to exercise external antennas and mobile setups. If you're in an urban area, the observation decks of tall buildings are also great for radio. You can operate from remote and unusual locations while rock climbing or hiking, for example, or just use the height of a building or hill to give you tremendous additional range!

Conducting Coverage Tests: Can You Hear Me Now?

Conducting coverage tests is a fun activity for a neighborhood group using personal FRS/GMRS or CB radios. The basic idea is to see how well you can communicate locally by radio. Start by making a list of important sites, such as fire stations or schools you're likely to want to communicate with in your neighborhood or within a certain range. Make a map of the locations and also those you think will be good or poor (that is, trickier) spots for radio communications.

Send someone to each location and see who can communicate with whom. All the participants should keep lists as they go. You'll probably have several rounds of communications as people move to new spots and each make efforts to communicate with everyone else in the group. Conclude the exercise with a cookout!

Conducting coverage tests is excellent preparation for emergencies.

Going on a Radio Scavenger Hunt

Radio scavenger hunts are great for kids and students using scanners or CB radios. What you should do is come up with a list of radio-related things that each contestant or team must gather. Some good examples of scavenger items include a signal from a policeman, a signal from a delivery truck, and a signal from someone whose name begins with the letter B, and so on. The more sophisticated the listeners, the better the game becomes.

Riding at a Radio Rodeo

A radio rodeo is a competition of your own design with any kind of radio the competitors can use. The sky is truly the limit. For example, you can compete to see who can receive a code number over the longest distance from a low-power FRS radio. You might also compete to see who can get a CB radio on the air from a car the quickest, or have a contest for the best handheld radio installation on a bicycle. Adult competitors can engage in road rally-style exercises using CBs and show off their workmanship. You can take cues from radio-control and robot-building hobbyists that get together and think up interesting competitions.

Creating Radio Scoreboards

You can use radio communications to set up a master scoreboard on Soccer Saturdays or Football Fridays or just announce scores. Within a single park with several fields, FRS/GMRS radios work fine. Around town and between parks, you need longer-range CB radios. It's fun to set up that out-of-town scoreboard for everyone to enjoy. Work with schools and let the public know in advance so they can listen in, too!

Going to a Hamfest

Part convention, part radio flea market, *hamfests* for ham radio enthusiasts, and similar gatherings (like the ones held by Citizens Band enthusiasts) offer a treasure trove of bargains. Hamfests aren't restricted to ham radio users.

In fact, non-hams can find numerous useful accessories, as well as many non-ham radios and supplies, computers, batteries, electronic supplies, and books.

Hamfests are surely great places to find stuff, but they are also great places to meet people and find radio clubs. If you're just starting out, building a network of likeminded hobbyists with more experience can help you take your skills to new levels. Find hamfests through radio clubs or at www.arrl.org/hamfests.html. This page is a courtesy offered by the American Radio Relay Association (ARRL), which is also commonly known as the National Association for Amateur Radio.

Weather Watching

I mention the SKYWARN weather watching program in Chapter 6. You don't have to join a formal weather watching program to keep an eye on the sky. Some professional broadcasters collect weather data from hams and other radio users at regular times each day for use on their news programs. For many people, keeping a weather log is a pleasurable hobby in and of itself. Sharing this data by radio takes two hobbies in new directions.

Wildlife Tracking

Radio-savvy volunteers are needed to help track migratory wildlife. If you love animals and are interested in contributing to animal research programs such as the programs described at members.aol.com/joemoell/owl.html, you should definitely join the cause. It's incredibly gratifying work, and very interesting. The Audubon Society (www.audubon.org) sponsors bird counting events and tours throughout the year. Radio technology helps coordinate these activities. The type of radio equipment varies with the need and is often provided or specified by the sponsor. Check with your local university or college biology department to find out about these programs.

Joining a Club

You can find lots of other folks out there using their radios for enjoyment, public service, or *emcomm* (emergency communication). Enter the search phrase **radio club** (or some variation of this term, such as **ham radio club** or **citizens band radio club**), along with the name of your city or state into an Internet search engine and you'll be amazed by the results. Also, magazines such as *QST* and *CQ* (for ham radio), *Popular Communications, Monitoring Times,* and Web publishers often have lots of leads. The emcomm organization REACT (www.reactintl.org) has councils and teams all over the U.S. that are great resources for new radio users. Multiply your radio enjoyment by sharing it with, and benefiting from, the expertise of others.

Appendix

Glossary

The following table defines some of the most common terms used in two-way radio and scanner operation. For words, abbreviations, and acronyms not present here, try one of the following online glossaries:

- **Monitoring Times:** www.grove-ent.com/mtglossary.html
- **ARRL:** www.arrl.org/qst/glossary.html
- **Strong Signals:** www.strongsignals.net/access/content/glossary.html

If you don't find the term you're looking for, try your radio's operating manual, the Web site of a radio manufacturer, or an Internet search engine. When using Google, you can enter *define: search term*. For example, if you can't remember what ac stands for, type **define: ac.**

ac (alternating current): An electric current that reverses direction at regular intervals. See also *dc.*

AF Gain: Receiver control that adjusts audio volume.

all call: A transmission intended to be received by all users at the same time.

amateur radio: A licensed radio service for individuals to experiment and train in radio techniques and perform emergency communications. Commonly referred to as *ham radio.*

ampere (amp, Ah): The unit of electrical current or flow.

amplitude: A measure that describes the magnitude (in volts, watts, amps, and so on) of a signal.

amplitude modulation (AM): A modulation technique that transmits information as variations in the strength or amplitude of a radio signal. See also *FM* and *modulation.*

analog system: A standard FM system whose signals can be received and understood directly. See also *digital system.*

antenna: A device for transmitting and receiving radio waves. The most common antenna types include the dipole and ground-plane vertical.

Two-Way Radios & Scanners For Dummies

APCO Project 25: The name of a trunked radio technology that transmits voice and data as digital radio signals.

attenuator: A circuit that reduces signal strength at the receiver input to prevent overload.

automatic frequency control (AFC): A receiver circuit that tracks a received signal. See also *phase-locked loop.*

automatic gain control (AGC): A receiver circuit that keeps output volume constant, independent of signal strength.

auxiliary connector: A connection for external equipment such as recorders or computers.

balun (balanced/unbalanced): A device to match an unbalanced feedline, such as coaxial cable, to a balanced antenna, such as a dipole, or a balanced feedline, such as twinlead, to an unbalanced antenna, such as a Yagi.

band: A range of frequencies. A generic term that refers to how much radio spectrum one signal consumes. See also *narrowband* and *wideband.*

bandwidth: The amount of frequency occupied by a radio signal.

bank: A segment of a scanner's memory.

base station: A radio station kept at a fixed location and used to contact mobile or portable stations. See also *station.*

beam antenna: An antenna that receives and transmits best in one direction.

broadcast: One-way transmissions intended for reception by a large number of receivers.

Business Radio Service (BRS): An FCC-licensed service for business and industrial users.

busy indicator: A light or display symbol on a radio that indicates that others are using the same radio channel. Useful when using CTCSS tones because you won't hear others not using the same tone. See also *CTCSS.*

call tone: An audible tone sent to alert others on the channel of an incoming transmission.

carrier: The part of an AM signal that is not modulated, but is used by the receiver to recover the sidebands. The carrier carries the sidebands which contain the modulating information (speech, music, data, and so on). See also *double sideband, sideband,* and *single sideband.*

channel: An assigned frequency for a radio transmission.

channel spacing: The frequency difference between two adjacent channels.

Citizens Band (CB): An unlicensed radio service operating at 27 MHz.

coaxial cable (coax): The most common cable used to carry radio signals. Has a woven or braided outer wire shield surrounding a cylinder of plastic insulation carrying a central wire. A plastic jacket covers the shield.

Continuous tone-coded sub-audible squelch: See *CTCSS*.

continuous wave (CW): A transmission on a single frequency. Because such a signal is turned on and off to generate Morse code, CW is often used as an abbreviation for Morse code. See also *Morse code.*

CTCSS (Continuous tone coded squelch system): A system that uses low-frequency audible tones to control whether a receiver output is turned on. Sometimes called *Continuous Tone Coded Sub-audible Squelch.*

dc (direct current): An electric current that flows continuously in one direction. See also *ac.*

DCS (digitally coded squelch): A squelch circuit that operates by sensing a code sent by the transmitter at the beginning of the transmission. See also *squelch.*

decibel (dB): A measure of the ratio of two signal amplitudes as a power of ten. For example, dB = 10 log (amplitude 1 ÷ amplitude 2).

demodulate: Recover information from a modulated radio signal.

digital system: A radio system in which voices are converted to a digital format before transmission. The voice data is usually combined with data and control signals to support identification, data exchange, and system operation.

dipole antenna: A ½-wavelength antenna made of wire, rod, or metal tubing with the feedline attached in the middle.

discone antenna: An omnidirectional antenna that works over a wide frequency range.

double sideband (DSB): A modulation technique that removes the carrier from an AM signal, but transmits both sidebands.

DTMF (Dual-Tone Multi-Frequency): The dialing and signaling system used by touch-tone phones. Pairs of audio frequencies used for signaling purposes.

duplex communications: Communications on a pair of frequencies, using one for transmitting and the other for receiving.

duty cycle: The percent of time equipment is on or in use.

duty factor: See *duty cycle.*

DX: Distant transmitter or distant station. See also *DXing* and *station.*

DXing: The reception of signals from distant transmitters.

EDACS (Enhanced Digital Access Communications System): A type of trunked radio system developed by Ericsson. See also *trunked radio system.*

element: A piece of an antenna that radiates or receives radio signals. A *driven element* is connected to a feedline.

encrypt: Render a digital data signal unintelligible by combining it with a code known only to the intended recipient. This is not the same as *scrambling,* an analog process, but the terms are often used interchangeably.

FCC (Federal Communications Commission): The government agency responsible for regulating telecommunications in the United States.

feedline: The cable used to connect a radio to an antenna.

frequency: The number of oscillations of a radio signal per second.

frequency modulation (FM): A modulation technique that transmits information as variations in the frequency of a radio signal. See also *AM* and *modulation.*

Family Radio Service (FRS): An unlicensed radio service introduced by the FCC for two-way radios operating between 462.5625 and 467.7125 MHz. See also *FRS/GMRS* and *GMRS.*

FRS/GMRS: A two-way radio package that bundles the Family Radio Service and General Mobile Radio Services. See also *FRS* and *GMRS.*

GHz (gigahertz, pronounced with a hard g): One billion hertz. See also *hertz.*

General Mobile Radio Service (GMRS): A licensed radio service introduced by the FCC for two-way radios operating between 462.550 and 467.725 MHz. See also *FRS* and *FRS/GMRS.*

Global Positioning System (GPS): A radiolocation system that uses satellites that transmit precisely timed codes. Terrestrial receivers compare the signals from several satellites to determine exact location.

group call: A transmission to a group of users.

ham radio: See *amateur radio.*

handheld (HT): Handheld radio.

high frequency (HF): Any signal transmitted at frequencies between 3 MHz and 30 MHz. See also *ultra high frequency* and *very high frequency.*

Hz (hertz): The unit of frequency. One cycle per second is one hertz.

ionosphere: The part of the earth's atmosphere that reflects radio signals back to the earth.

keypad lock: A function on a radio that disables the keypad to prevent accidentally changing the radios operation or programming.

kHz (kilohertz): One thousand hertz. See also _hertz._

liquid crystal display (LCD): A screen for displaying text and graphics.

Li-Ion (Lithium-Ion): A type of rechargeable battery pack or cell.

logic trunked radio (LTR): A type of trunked radio system that uses logical channels and transmits both voice and control information on the same frequency.

log-periodic antenna: A beam antenna that works over a wide frequency range.

low frequency (LF): See _longwave._

longwave (LW): Radio signals transmitted below 300 kHz. LW is often used to refer to all radio signals below 540 kHz.

lower sideband (LSB): The sideband lower in frequency than the carrier. See also _sideband_.

manual channel access: The ability to tune directly from one channel to another, skipping the channels in between.

medium wave (MW): Radio signals from 300 kHz to 3,000 kHz. MW is often used to refer to any radio signal in the AM broadcast band (540 to 1,700 kHz).

MHz (megahertz): One million hertz or cycles per second. See also _hertz._

mobile radio: A radio that's permanently mounted in a vehicle or that operates while in motion.

mode: A type of signal, such as AM, CW, FM, NFM, SSB, FSK, RTTY, and so on. See also _modulation_.

modulation: The process of combining information with a radio signal so that it can be transmitted. The modulation is named for the aspect of the signal that is varied in order to represent the information, such as its amplitude (amplitude modulation, or AM) or frequency (frequency modulation, or FM).

monitor: To listen to a channel. A monitor button temporarily opens the squelch system to listen to any signal present.

Morse code: A radio signal turned on and off in the pattern of Morse code characters. See also *continuous wave.*

multicast: A radio system in which transmitters use a set of different frequencies and the receivers tune to the channel with best reception. See also *simulcast.*

multisite: A radio system that extends coverage beyond single-site systems by using two or more repeaters or transmitters linked together.

Multi-Use Radio Service (MURS): An unlicensed two-way radio service established by the FCC. MURS radios operate at frequencies between 151.820 MHz and 164.600 MHz.

narrowband: A range of frequencies that occupies 12.5-kHz channels. See also *band* and *wideband.*

Narrowband FM (NFM): A type of FM signal with less frequency variation than the maximum allowed for the service.

NBFM: See *NFM.*

NiCd (nickel cadmium): A type of rechargeable battery pack or cell using nickel and cadmium.

NiMH (nickel metal hydride): A type of rechargeable battery pack or cell using nickel and a metal hydride.

NOAA: National Oceanic and Atmospheric Administration.

noise blanker (NB): A receiver circuit that removes noise pulses or static crashes from the output.

notch filter: A filter that removes a very narrow range of frequencies from a received signal. Notch filters can either operate on the radio frequency itself *(RF notch)* or on an audio signal *(audio notch).*

NWS: National Weather Service.

ohm: the units of electrical resistance. See also *resistance.*

PL (Private Line): See *CTCSS.*

phase-locked loop (PLL): A tuning circuit that locks onto and tracks a desired signal by duplicating its phase (a measure of the signal's change in frequency). The PLL is used to both generate and receive FM signals.

Private Land Mobile Radio (PLMR): A service created by the FCC for industrial and business users whose operators are mostly mobile or portable, such as taxi and delivery services.

PMR 446 (Private Mobile Radio service): A European service, similar to FRS in the U.S., that operates at 446 MHz.

portable: Radio that is intended to be carried while in use.

preamp (preamplifier): A circuit used to increase a receiver's sensitivity.

priority channel: A channel that is scanned more frequently than non-priority channels.

Push to Talk (PTT): A switch on a radio that's used to turn on or key the transmitter.

QRM: Abbreviation signifying interference from other transmitted signals.

QRN: Abbreviation signifying interference from atmospheric or other non-transmitted sources, such as motors, power lines, and so on.

QSL: Abbreviation meaning "received and understood." Also refers to a letter or postcard confirming or claiming reception of a signal.

QSO: Abbreviation signifying that a radio contact has been made.

receiver: An electronic device intended to receive radio signals by using an antenna. See also **_transceiver._**

repeater: A relay station that listens on one frequency and retransmits the signal on another frequency.

reset: To return a radio to its original factory settings.

resistance: the ratio of electrical current to applied voltage.

RF (radio frequency): Any signal high enough in frequency so as to be inaudible. Typically any signal greater than 20 kHz.

RF gain: The control that varies receiver sensitivity. See also **_sensitivity._**

rubber duck: Flexible, shortened, handheld radio antenna covered with a rubberized coating.

S-Meter: Signal strength meter, shows strength of received signal.

scan: To repeatedly tune through a programmed sequence of channels.

scan delay: A specified amount of time a scanner remains on a busy channel before resuming the scan.

scan rate: The number of channels a scanner can tune per second.

scanner: A radio that is designed to tune quickly and/or automatically across a sequence of channels. See also **_receiver._**

scramble: To change an analog voice or data signal to an altered form that cannot be received by a listener without knowledge of the alteration method.

search: See *sweep.*

selectivity: The ability of a receiver to discriminate between a desired signal and other signals on nearby channels or frequencies.

sensitivity: The ability of a receiver to detect and recover weak signals.

shared system: A system that supports small businesses that need wide coverage, but which can't afford to purchase a repeater. Vendors often maintain and rent a radio system to many different subscribing users.

shortwave (SW): Radio signals transmitted at frequencies between 2 and 30 MHz. Also called **world band**.

sideband: The part of an AM signal that actually contains the modulating information. An AM signal contains a pair of sidebands, one higher in frequency than the carrier, and one lower in frequency. See also *AM.*

simplex: Communications that occur on one frequency, alternating between transmitting and receiving.

simulcast: A system that provides wider coverage with multiple transmitters, all of which broadcast the same signal simultaneously on a common frequency. See also *multicast.*

single sideband (SSB): A modulation technique that removes one sideband and the carrier and transmits only the remaining sideband. See also *sideband.*

single-site: A system that has just one base station or repeater, covering an area centered on that transmitter. Operation can be dispatcher controlled or one to one. See also *multisite.*

subscribed system: See *shared system.*

Specific Area Message Encoding (SAME): A protocol that broadcasts messages targeted to a single area. It is used by NOAA weather stations to encode and transmit emergency and severe weather information.

squelch: A circuit that mutes the receiver audio output until the strength of the signal exceeds a desired level.

SSB: See *single sideband*.

station: A radio system used to transmit and receive radio signals.

step size: The size of frequency changes as receiver is tuned between channels. For example, when scanning the GMRS channels, the step size would be set to 25 kHz.

Store and forward: A radio system that records a signal when it is present and retransmits it as soon as it disappears.

synchronous detection: A receiver circuit or mode that reduces the effects of fading on AM signals.

sweep: To tune continuously across a range of frequencies.

talk group: A group of trunked radio system users that communicate with each other.

telescopic antenna: A whip antenna that can be lengthened or shortened to be stored or to vary reception on different frequencies.

time-out timer: A circuit that turns off the transmitter after a preset length of time.

tone squelch: See *CTCSS.*

transceiver: transmitter/receiver. A radio that can both transmit and receive radio signals.

transmitter: An electronic device that generates radio signals intended to be radiated by an antenna.

trunked radio system (TRS): A radio system that allows several users to share the same set of channels under the control of a central master computer.

twin lead cable: A type of feedline consisting of two parallel wires separated and covered by plastic insulation.

two-way radio: Radio intended to allow communications between two points, with signals being transmitted in both directions.

ultra high frequency (UHF): Signals sent at frequencies that range from 300 MHz to 3 GHz. See also *high frequency* and *very high frequency.*

upper sideband (USB): The sideband higher in frequency than the signal's carrier. See also *sideband*.

variable frequency oscillator (VFO): The device that controls the frequency of a receiver or transmitter; a mode of tuning where the signal frequency is varied continuously or in steps smaller than the usual channel spacing at that frequency.

very high frequency (VHF): Frequencies from 30 MHz to 300 MHz. See also *high frequency* and *ultra high frequency.*

volt: The unit of electrical potential or force.

volt ohm meter (VOM): An instrument that can measure both voltage and resistance. A VOM often measures current and other electrical parameters.

voting: When multiple satellite receivers are added to increase the receive range of the system, a comparator circuit selects the best signal from the receivers to be transferred to the repeater or dispatcher.

VOX (voice-operated transmit): A circuit that senses the user's voice to turn on the transmitter. See also *DX.*

VOX sensitivity: The level of microphone input that causes the VOX circuit to turn on the transmitter.

watt: The unit of power.

wavelength: The distance a radio wave travels while making one complete oscillation. See also *frequency.*

weather channel (WX): A channel provided by the NOAA's National Weather Service that broadcasts weather updates and alerts.

weather alert: Automatic detection of emergency or severe weather alerts from NOAA weather stations.

whip antenna: A self-supporting antenna consisting of a single, flexible metal rod or wire, sometimes contained in an insulating or protective tube of plastic or fiberglass. See also *telescopic antenna.*

wideband: A range of frequencies that use 25-kHz channels.

Wideband FM (WFM): A type of FM signal whose frequency variations are the widest allowed for the service.

world band: See *shortwave.*

Yagi: A type of beam antenna named for its inventor, Dr. Yagi. The Yagi uses straight elements arranged so as to reflect or direct signals radiated from a single driven element. See also *element.*

Index

Printed in Great Britain by
Amazon.co.uk, Ltd.,
Marston Gate.